Antennas

F. R. Connor
Ph.D., M.Sc., B.Sc.(Eng.) Hons, A.C.G.I.,
C.Eng., M.I.E.E., M.I.E.R.E., M.Inst.P.

R.F. Cheeney
1976 January.

Edward Arnold

First published 1972 by Edward Arnold (Publishers) Ltd., 25 Hill Street,
London, WIX 8LL

ISBN 0 7131 3279 5

Photoset by The Universities Press, Belfast
and printed in Great Britain by The Pitman Press, Bath.

Preface

This is an introductory book on the important topic of *Antennas*. Various types of antennas are used in practice, each with its own particular characteristics, and the book endeavours to present the basic ideas in a concise and coherent manner. Moreover, to assist in the assimilation of these basic ideas, many worked examples from past examination papers are provided to illustrate clearly the application of the fundamental theory.

The book begins with a review of features common to many antennas and a brief analysis is made of some simple types of single antenna structures. To obtain greater directivity, various forms of arrays are considered in the following chapters and a description of some important microwave antennas is included. The book ends with a chapter on the propagation of electromagnetic waves from low frequencies to very high frequencies.

This book will be found useful by students preparing for London University examinations, degrees of the Council of National Academic Awards, examinations of the Council of Engineering Institutions and for other qualifications such as Higher National Certificates, Higher National Diploma and certain examinations of the City and Guilds of London Institute. It will also be useful to practising engineers in Industry who require a ready source of basic knowledge to help them in their applied work.

Acknowledgements

The author sincerely wishes to thank the Senate of the University of London and the Council of Engineering Institutions for permission to include questions from past examinations papers. The solutions and answers provided are his own and he accepts full responsibility for them.

Sincere thanks are also due to the City and Guilds of London Institute for permission to include questions from past examination papers. The Institute is in no way responsible for the solutions and answers provided.

Finally, the author wishes to thank the publishers for various useful suggestions and will be grateful to his readers for drawing his attention to any errors which may have occurred.

F.R.C.

Books in this series by the same author:

1. Signals
2. Networks
3. Wave transmission

In preparation:

5. Modulation
6. Noise

Contents

Symbols used in the book

λ wavelength
ω angular frequency
σ $\begin{cases} \text{conductivity} \\ \text{echoing area} \end{cases}$
φ phase difference
ρ $\begin{cases} \text{reflection coefficient} \\ \text{radius of curvature} \end{cases}$
ν collision frequency
c velocity of light
k Boltzmann's constant
f_c critical frequency
n $\begin{cases} \text{any integer} \\ \text{refractive index} \end{cases}$
A effective aperture area
$\mathbf{A}$ magnetic vector potential
B bandwidth
G power gain
J current density
T absolute temperature

Abbreviations used in the book

C.E.I. Part 2	The Council of Engineering Institutions examination in Communication Engineering, Part 2.
C. & G.	The City and Guilds of London Institute.
L.U., B.Sc(Eng) Tels.	University of London, B.Sc(Eng) examination in Telecommunications, Part 3.

1
Introduction

An antenna or aerial is a device for radiating or receiving electromagnetic waves and there is little fundamental difference between transmitting and receiving antennas, since very often the same antenna is used for both purposes as in radar. Antennas are therefore associated with certain definite properties which are common to them all.

Since antennas play a very important part in a communication system, it is essential to have a thorough understanding of the principles on which they perform their task efficiently. Hence, this has led to the design and development of a great many different types of antennas, each suited to a particular set of applications. Whereas some antennas can play a versatile role such as the simple piece of wire thrown out of the window, for the best performance, the most suitable type of antenna should be used under proper operating conditions.

An understanding of the basic properties underlying antennas will be considered first before undertaking a description and analysis of the most important types of antennas. A further point to be considered is the fact that in the past, antennas have been treated somewhat separately from the system in which they are used. But in recent years, antennas are being treated as an essential part of the system and are designed alongside the system. This is because it is essential to know about the system performance as a *whole*, rather than that of the antenna *alone*. For example, in a satellite ground station, the noise temperature of the *whole* system is a criterion of performance when it comes to receiving weak satellite signals and not just the noise temperature of the antenna.

1.1 Polarisation

An electromagnetic wave launched from an antenna may be vertically or horizontally polarised. In the former case, the E vector is vertical and requires a vertical antenna to launch it. Alternatively, if the E vector is horizontal, the wave is horizontally polarised and requires a horizontal antenna to launch it. Sometimes, circular polarisation is used, which is a combination of vertical and horizontal polarisation. Electromagnetic waves are usually vertically polarised though the other types of polarisation may be used for specific purposes on occasions.

1.2   Polar diagram

By far, the most important property of an antenna is its radiation pattern or polar diagram. In the case of a transmitting antenna, this is a plot of the power or field strength radiated by the antenna in different angular directions. A plot may be obtained for the vertical or horizontal planes and are called the vertical and horizontal polar patterns respectively. Incidentally, the same patterns apply whether the antenna is transmitting or receiving radiation because of the principle of reciprocity.*

1.3   Antenna gain

As a consequence of its polar pattern, power radiated by an antenna may be concentrated in a particular direction. This directivity is usually expressed in terms of its power gain G which is defined *in a particular direction* as

$$G = \frac{\text{power radiated by an antenna}}{\text{power radiated by reference antenna}}$$

The input powers to both antennas is the same and the reference antenna usually chosen is the isotrope or point source which radiates equally in all directions or occasionally, it may be a dipole.

The power radiated by the antenna is slightly less than the input power because of losses in the antenna. An alternative definition which assumes 100% efficiency is defined as the directive gain D such that $G = \eta D$ where $\eta < 1$ and so G is slightly less than D. In practice, the power gain G is commonly used and it is expressed as a pure number or in dB.

1.4   Radiation resistance

An important property of a transmitting antenna is its radiation resistance which is associated with the power radiated by the antenna. If I is the r.m.s. antenna current and R_r is its radiation resistance, then the power radiated is $I^2 R_r$ watts where R_r is a *fictitious* resistance which accounts for the radiated power somewhat like a circuit resistance which dissipates heat. The radiation resistance should be large as the greater R_r is, the greater the power radiated by the antenna.

In contrast, for a receiving antenna, its input impedance is important.

* See Section 1.8.

The input impedance is defined as the ratio of voltage to current at its input and it must be generally matched to the connecting line or cable. The input impedance may or may not equal its radiation resistance, though very often it does, as for example, a centre-fed dipole has an input impedance of about 75 Ω which equals its radiation resistance of 75 Ω. In most cases, R_r may be calculated or it can be determined experimentally.

1.5 Effective length

An antenna with a non-uniform distribution of current over its length l can be considered as having a shorter effective length l_e over which the current is assumed to be uniform and equal to its peak value. The relationship between l_e and l is given by

$$\frac{l_e}{l} = \frac{\text{area under non-uniform current distribution}}{\text{area under uniform peak current distribution}}$$

1.6 Effective aperture

The power received by an antenna can be associated with a collecting area. Every antenna may be considered to have such a collecting area which is called its effective aperture A. If P_d is the power density at the antenna and P_r is the received power, then

$$P_r = P_d A \text{ watts}$$

or

$$A = P_r/P_d \text{ m}^2$$

In Appendix A it is shown that for an antenna with power gain G, the effective aperture A at the operating wavelength λ is given by

$$A = G\lambda^2/4\pi \text{ m}^2$$

1.7 Power transfer

For the maximum transfer of power from a receiving antenna to its receiver, the impedance of the antenna should be matched to the input impedance of the receiver, in accordance with the maximum power transfer theorem.[1] As the antenna impedance is normally resistive, this means that the input impedance of the receiver is also resistive. Fig. 1 shows an antenna of impedance Z_{in} connected to a receiver with the input terminated in a resistor R_g.

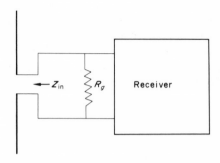

Fig. 1

If V is the induced r.m.s. voltage in the antenna, then the current i is given by

$$i = \frac{V}{Z_{in} + R_g}$$

The power transferred to the receiver is

$$P_r = i^2 R_g = \frac{V^2 R_g}{(Z_{in} + R_g)^2}$$

For a maximum, $$\frac{dP_r}{dR_g} = 0$$

or $$\frac{(Z_{in} + R_g)^2 - 2R_g(Z_{in} + R_g)}{(Z_{in} + R_g)^4} = 0$$

Hence $$Z_{in} + R_g = 2R_g$$

or $$Z_{in} = R_g$$

and $$P_r(\text{max}) = \frac{V^2 R_g}{4R_g^2} = \frac{V^2}{4R_g}$$

1.8 Reciprocity

The properties of a transmitting antenna are very similar to those of a receiving antenna because of the theorem of reciprocity which states that if an e.m.f. is applied to the terminals of a circuit A and produces a current in another circuit B, then the same e.m.f. applied to terminals B, will produce the same current at the terminals of circuit A.

Hence, because of this reciprocal result, the polar diagram of a receiving antenna is similar to that when it is transmitting. This applies equally well to the power gain of the antenna in either mode. Finally, it also implies that for a *given arrangement*, the antenna impedance is the same, whether it is transmitting or receiving radiation.

2
Elementary antennas

A very elementary yet important theoretical antenna is the point source radiator or isotrope. It serves as a basis for comparison of many types of antennas whose performance is best expressed in terms of such a basic radiator.

2.1 The isotrope

An isotrope or point source radiator radiates energy equally in all directions. Hence, its radiation is isotropic and the radiation pattern in any plane is a circle.

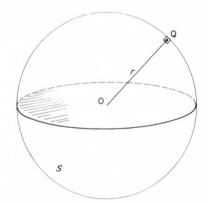

Fig. 2

Consider such an isotrope at point 0 (Fig. 2) fed with power P watts. The power flows outwards from 0 and at any time t it must flow through the spherical surface S of radius r. Hence, the power density P_d at point Q is given by

$$P_d = P/4\pi r^2 \text{ watts/m}^2$$

By Poynting's theorem,[2] the power density P_d is related to the E and H vectors by the equation

$$P_d = \mathbf{E} \times \mathbf{H} \text{ watts/m}^2$$

or $$|P_d| = E \cdot H = E^2/120\pi$$

since $$E/H = 120\pi$$

Hence $$\frac{E^2}{120\pi} = P/4\pi r^2$$

or $$E = \frac{\sqrt{30P}}{r} \text{ V/m}$$

where E and H are r.m.s. values when P is the average power.

The isotrope may be considered as a standard reference antenna with power gain $G = 1$. If any other practical antenna with a power gain G were placed at 0, then the power received at Q would be increased G times to $G . P$ watts. Hence, the field strength at Q will be increased to

$$E = \frac{\sqrt{30GP}}{r} \text{ V/m}$$

Radar equation

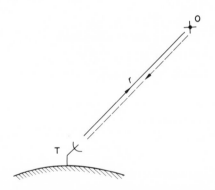

Fig. 3

Suppose a radar transmitter at T in Fig. 3 has an antenna gain G_T and is transmitting a power P_T. The power density P_d at an object located at 0 distant r is

$$P_d = \frac{P_T . G_T}{4\pi r^2} \text{ watts/m}^2$$

The target area of the object at 0 is defined as σ m^2 such that the power reflected back *isotropically* is $P_d . \sigma$. It therefore produces a power density P'_d at the transmitter where

$$P'_d = \frac{P_T \cdot G_T \sigma}{(4\pi r^2)^2} \text{ watts/m}^2$$

If a receiving antenna at T has a gain G_R and effective area A, then the received power is

$$P_r = P'_d \cdot A = \frac{P_T \cdot G_T \cdot G_R \sigma}{(4\pi r^2)^2} \times \frac{\lambda^2}{4\pi}$$

or

$$\frac{P_r}{P_T} = \frac{G_T \cdot G_R \sigma \cdot \lambda^2}{(4\pi)^3 \cdot r^4}$$

This is the basic radar equation relating transmitted and received powers. If the minimum detectable signal-to-noise ratio at the receiver is S/N, then $P_r = S$ and $N = kTB$ where k is Boltzmann's constant, T is the absolute temperature and B is the bandwidth of the system. The radar range r is then given by

$$r^4 = \frac{P_T \cdot G_T \cdot G_R \cdot \sigma \cdot \lambda^2}{(4\pi)^3 kTB \cdot (S/N)}$$

or

$$r = \left[\frac{P_T \cdot G_T \cdot G_R \cdot \sigma \cdot \lambda^2}{(4\pi)^3 kTB \cdot (S/N)} \right]^{1/4}$$

The range depends on the fourth root of P_T and so accounts for the large powers used in radar systems.

EXAMPLE 1
Calculate the minimum peak transmitter power needed in a pulsed radar required to detect a target of 10 m² echoing area at a range of 120 km, given the following system parameters

operating frequency	1·3 GHz
receiver sensitivity	− 105 dBm (dB rel. to 1 mW)
aerial gain	34 dB
atmospheric attenuation	0·008 dB/km

How would the pulse length and pulse repetition frequency of the radar be determined?   (C.E.I. Comm. Eng. Oct. 1968)

Solution

From the previous analysis we obtained

$$\frac{P_T}{P_r} = \frac{(4\pi)^3 \cdot r^4}{G_T \cdot G_R \cdot \sigma \cdot \lambda^2}$$

or

$$P_T = \frac{(4\pi)^3 \cdot r^4 \cdot P_r}{G_T \cdot G_R \cdot \sigma \cdot \lambda^2}$$

Due to atmospheric attenuation over the *two-way* distance there is a path loss given by

$$\text{Path loss} = 2 \times 120 \times 0{\cdot}008 = 1{\cdot}92\,\text{dB}$$

Hence, the transmitted power P_T must be increased by $1{\cdot}92\,\text{dB}$ or $1{\cdot}556$ times.

Hence

$$P_T = \frac{(4\pi)^3 r^4 P_r (1{\cdot}556)}{G_T G_R \sigma \lambda^2}$$

Here

$$r = 120\,\text{km} = 1{\cdot}2 \times 10^5\,\text{m}$$

$$G_T = G_R = 34\,\text{dB} = 2512$$

$$\sigma = 10\,\text{m}^2$$

Also

$$\text{f}\,.\,\lambda = 3 \times 10^8$$

or

$$\lambda = \frac{3 \times 10^8}{1{\cdot}3 \times 10^9} = 0{\cdot}23\,\text{m}$$

Since the receiver sensitivity is $-105\,\text{dBm}$,

$$P_R = -105\,\text{dBm} = -135\,\text{dBW} = 10^{-13.5}\,\text{Watts}$$

Hence

$$P_T = \frac{(4\pi)^3 \times (1{\cdot}2 \times 10^5)^4 \times 10^{-13.5} \times 1{\cdot}556}{(2512)^2 \times 10 \times (0{\cdot}23)^2}$$

or

$$P_T = 6\,\text{kW}$$

In the last part of the question, the pulse length and pulse repetition frequency determine the minimum range and maximum range of the radar respectively.*

* See F. R. Connor, SIGNALS, Edward Arnold (1972).

2.2   Hertzian dipole

A simple practical antenna is the doublet or Hertzian dipole shown in Fig. 4. It is a very short length of wire over which the current distribution can be assumed to be uniform. Maxwell's equations show that such an antenna when energised by a high-frequency current is associated with an induction field which decreases inversely as the *square* of the distance and a radiation field, which decreases inversely as *distance* only. The latter is still measurable at large distances from the doublet and is the well-known radiation field used in radio communications.

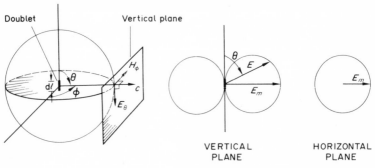

Fig. 4

The magnetic field at a distance r is shown in Appendix B to be

$$H_\varphi = \frac{I_0 \cdot dl \cdot \sin\theta}{4\pi}\left[\frac{je^{j\omega(t-r/c)} \cdot \omega}{cr} + \frac{e^{j\omega(t-r/c)}}{r^2}\right]$$

The first term is the radiation field and the second term is the induction field. Hence, for the radiation field we obtain

$$H_\varphi = \frac{jI_0 \cdot e^{j\omega(t-r/c)} \cdot dl \cdot \sin\theta}{4\pi}(\omega/cr) = \frac{jI_0 \cdot e^{j\omega(t-r/c)} \cdot dl \sin\theta}{2\lambda \cdot r}$$

Also, from Maxwell's equations we have

$$\frac{E_\theta}{H_\varphi} = 120\pi$$

or

$$E_\theta = \frac{j \cdot 60\pi \, dl \cdot \sin\theta \cdot I_0 \cdot e^{j\omega(t-r/c)}}{\lambda \cdot r}$$

If the r.m.s. current in the doublet is I and the r.m.s. value of E_θ is E, then

$$E = \frac{60\pi I \cdot \mathrm{d}l \sin\theta}{\lambda \cdot r} \text{ V/m}$$

or
$$E = E_m \sin\theta$$

where
$$E_m = \frac{60\pi I \, \mathrm{d}l}{\lambda \cdot r}$$

This result is for a free space doublet and the polar diagram in the vertical plane is a 'figure of eight' and in the horizontal plane, it is a circle, as shown in Fig. 4. The electric field is directional in the vertical plane but is omnidirectional in the horizontal plane. This gives the doublet a power gain $G = 1\cdot5$ or $1\cdot76$ dB relative to an isotrope.

Radiation resistance

The power radiated by an antenna is given by the Poynting[2] vector theorem $\mathscr{P} = \mathbf{E} \times \mathbf{H}$ watts/m^2. If the radiation resistance of the doublet is R_r, then its value may be obtained by equating the power radiated I^2R_r to $\mathscr{P}$.

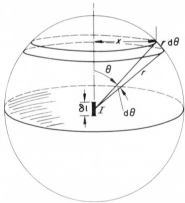

Fig. 5

Consider the doublet at the centre of a sphere of radius r and energised by an r.m.s. current I as in Fig. 5. The area of a sector of radius x and width $r\,\mathrm{d}\theta$ is $2\pi x \cdot r\,\mathrm{d}\theta = 2\pi r^2 \cdot \sin\theta\,\mathrm{d}\theta$ since $x = r\sin\theta$. The average power flow $\mathrm{d}P$ through this sector is

$$\mathrm{d}P = \frac{E^2}{120\pi} \cdot (2\pi r^2 \sin\theta\,\mathrm{d}\theta)$$

where E is the r.m.s. value

or
$$dP = \frac{E^2 r^2}{60} \sin\theta \, d\theta$$

Now
$$E = \frac{60\pi I \cdot dl \cdot \sin\theta}{\lambda \cdot r} \quad \text{(using r.m.s. values)}$$

Hence
$$dP = 60\left(\frac{\pi I \cdot dl}{\lambda}\right)^2 \sin^3\theta \, d\theta$$

The total power flow P through the sphere is given by

$$P = \int_0^\pi dP = \int_0^\pi 60\left(\frac{\pi I \, dl}{\lambda}\right)^2 \sin^3\theta \, d\theta$$

or
$$P = 60\left(\frac{\pi I \, dl}{\lambda}\right)^2 \int_0^\pi \sin^3\theta \, d\theta$$

Since the integral equals 4/3 we obtain

$$P = 80\pi^2 I^2 \left(\frac{dl}{\lambda}\right)^2 \text{ watts}$$

If the radiation resistance is R_r and the r.m.s. antenna current is I, then

$$I^2 R_r = P = 80\pi^2 I^2 \left(\frac{dl}{\lambda}\right)^2$$

or
$$R_r = 80\pi^2 \left(\frac{dl}{\lambda}\right)^2 \text{ ohms}$$

Comment
The results obtained above for a vertical doublet also apply to a horizontal doublet, if the horizontal and vertical polar diagrams are interchanged.

2.3   Short antenna

The nearest practical approach to a doublet is an antenna of very short length i.e. $dl < \lambda/10$. In this case, a uniform current distribution may still be assumed but for lengths between $\lambda/10$ to $\lambda/4$, a linear distribution may be used if one end is grounded.

EXAMPLE 2
The phasor representing the electric field strength radiated by a wire of
length dl and carrying a current represented by the phasor I, at distance r
from the centre of the wire is

$$E = \frac{60\pi I \, dl \sin \theta}{\lambda \, . \, r} . e^{-2\pi jr/\lambda}$$

where θ defines the direction of the point at which E is measured with
respect to the wire axis and λ is the free space wave-length.

A dipole of length 1 m is used at a wavelength of 10 m. Calculate the
radiation resistance of the dipole. What would be the nature of the
reactive component of the dipole impedance?

(C.E.I. Part 2, Comm. Eng. 1968)

Solution
Here
$$dl = 1 \text{ metre}$$
$$\lambda = 10 \text{ metres}$$

Since $dl = \lambda/10$, the dipole is very short and the derivation of R_r given
in Section 2.2 is valid where

$$R_r = 80\pi^2 \left(\frac{dl}{\lambda}\right)^2$$

Hence
$$R_r = 80\pi^2 \times \frac{1}{10^2} = 0.8\pi^2$$

or
$$R_r = 7.9 \, \Omega$$

A short antenna may be regarded approximately as a series tuned
circuit. Such a circuit is essentially capacitive below resonance and
inductive above resonance. Since the wavelength is 10 m which is ten
times the resonant wavelength of the dipole i.e. the operating frequency is
relatively low compared to the resonant frequency, the reactive component
of the dipole impedance is essentially capacitive.

2.4 Loop antenna[3]

A vertical loop antenna may be found by combining the radiation from
four doublets, two being vertical and two horizontal as in Fig. 6.

In the XY plane the fields due to doublets 3 and 4 are equal and opposite
and so cancel out, whereas the fields E_θ of doublets 1 and 2 combine to
give a resultant field E_R.

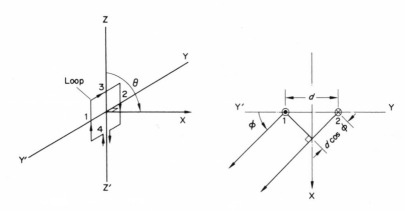

Fig. 6

The value of E_θ in the XY plane for a single doublet is obtained from Section 2.2 with $\theta = 90°$. Hence

$$E_\theta = \frac{j60\pi I_0 \, dl \cdot e^{j\omega(t-r/c)}}{\lambda \cdot r}$$

The resultant field E_R is given by the phasor sum of the fields of the two doublets, with a path difference $d \cos \varphi$. If ψ is the corresponding phase difference then

$$\psi = \frac{2\pi d \cos \varphi}{\lambda} + 180°$$

where $+180°$ is due to the currents in doublets 1 and 2 being equal but opposite. Hence, the resultant field E_R is given by

$$E_R = 2E_\theta \cos \psi/2 = 2E_\theta \cos\left(\frac{\pi d \cos \varphi}{\lambda} + 90°\right)$$

$$= -2E_\theta \sin\left(\frac{\pi d \cos \varphi}{\lambda}\right)$$

or $$E_R = -2jE_\theta\left(\frac{\pi d \sin \varphi}{\lambda}\right)$$

since the quantity in brackets is small.

Substituting for E_θ then yields

$$E_R = 120\pi I_0 \, dl e^{j\omega(t-r/c)} \cdot \left(\frac{\pi d \sin \varphi}{\lambda^2 r}\right)$$

or

$$E_R = \frac{120\pi^2 I_0 A e^{j\omega(t-r/c)} \cdot \sin \varphi}{\lambda^2 r}$$

where $A = dl$. d is the area of the loop and E_R is independent of θ as the vertical polar pattern is a circle. The power radiated from the loop is shown in Appendix C to be

$$P = 160\pi^4 I_0^2 \left(\frac{A}{\lambda^2}\right)^2$$

If R_r is the radiation resistance of the loop and I_0 is the peak current in the loop, then

$$\tfrac{1}{2} I_0^2 R_r = 160\pi^4 I_0^2 \left(\frac{A}{\lambda^2}\right)^2$$

or

$$R_r = 320\pi^4 \left(\frac{A}{\lambda^2}\right)^2 \text{ ohms}$$

EXAMPLE 3

An earth station is receiving transmissions from a space research satellite on a frequency of 136 MHz. The satellite is at a range of 500 km and its transmitter supplies 0·5 W into an aerial having a gain of 3 dB with reference to an isotropic aerial. Assuming free space propagation, and taking the impedance of free space as 120π ohms, calculate
(a) the power flux density in watts/m².
(b) the field strength in $\mu V/m$ at the earth station.

If the aerial at the earth station has a gain of 20 dB with reference to an isotropic aerial, what is the signal power received? (The effective absorbing area of an isotropic aerial is $\lambda^2/4\pi$.)

(C & G Comm. Radio C, June 1968)

Solution
(a)

$$|P_d| = E \cdot H = E^2/120\pi$$

where

$$E = \frac{\sqrt{30GP}}{r} \quad \text{and} \quad G = 2$$

Hence $$P_d = \frac{30GP}{r^2 \cdot 120\pi} = \frac{30 \times 2 \times 1/2}{(500 \times 10^3)^2 \cdot 120\pi}$$

or $$P_d = 0.318 \times 10^{-12} \text{ W/m}^2$$

(b) $$E = \frac{\sqrt{30GP}}{r} = \frac{\sqrt{30 \cdot 2 \cdot 1/2}}{500 \times 10^3} = \frac{\sqrt{30}}{500 \times 10^3}$$

or $$E = 11 \,\mu\text{V/m}$$

If P_r is the received power at the earth station then

$$P_r = P_d \cdot A \cdot G_R = 0.318 \times 10^{-12} \times \frac{\lambda^2}{4\pi} \times 100$$

Since $$\lambda = \frac{3 \times 10^8}{136 \times 10^6} = 2.2 \text{ m}$$

Hence $$P_r = 0.318 \times 10^{-12} \times (2.2)^2 \times 100/4\pi$$

or $$P_r = 12.3 \times 10^{-12} \text{ watts}$$

EXAMPLE 4

Explain what is meant by the terms power gain and radiation resistance, as applied to aerials.

Deduce expressions for the electric and magnetic field intensities at a large distance r in the direction of maximum radiation from an aerial having a power gain G in that direction, a radiation resistance R_r and an excitation current I.

A transmitting aerial has a radiation resistance of 50 Ω and a power gain of 20 dB in the direction of a receiver 40 miles distant. With the aerial supplied with a current 0·5 A, determine the intensity (W/m²) and electric field strength at the receiver. If the receiving aerial has an effective length of 1·5 m and its radiation resistance is 75 Ω, determine the maximum power available to the receiver and the overall transmission loss in dB.

(L.U. B.Sc(Eng) Tels. Pt. 3, 1967)

Solution

The expressions power gain and radiation resistance are defined in Sections 1.3 and 1.4.

PROBLEM

The total power radiated by an antenna energised by a current I and with a radiation resistance R_r is $I^2 R_r$. This power flows out through a surface area $4\pi r^2$ at a distance r. Hence, the power density P_d at a distance r is

$$P_d = I^2 R_r / 4\pi r^2$$

If the electric field intensity at a distance r is E, then

$$P_d = E . H = E^2/120\pi$$

and for an antenna of gain G in the direction of maximum radiation, the radiated power is increased G times. Hence

$$G . \frac{I^2 R_r}{4\pi r^2} = E^2/120\pi$$

giving $\qquad E^2 = 30G . I^2 . R_r / r^2$

or $\qquad E = \sqrt{30G . I^2 . R_r}/r$

and $\qquad H = E/120\pi = \sqrt{30G . I^2 . R_r}/120\pi r$

The field intensity $P_d = E . H$ watts/m². Hence

$$P_d = \frac{\sqrt{30GI^2 . R_r}}{r} \times \frac{\sqrt{30GI^2 R_r}}{120\pi r}$$

$$= \frac{GI^2 R_r}{4\pi r^2}$$

$$= \frac{100 \times \frac{1}{4} \times 50}{4\pi \times (40 \times 1600)^2}$$

or $\qquad P_d = 0{\cdot}0243 \ \mu\text{W/m}^2$

Also $\qquad E = \frac{\sqrt{30GI^2 . R_r}}{r}$

$$= \frac{(30 \times 100 \times \frac{1}{4} \times 50)^{1/2}}{40 \times 1600}$$

or $\qquad E = 3{\cdot}02 \ \text{mV/m}$

If V is the voltage induced in the receiving antenna, then

$$V = E . l = 3{\cdot}02 \times 10^{-3} \times \tfrac{3}{2} = 4{\cdot}53 \ \text{mV}$$

The maximum power available to a receiver under matched conditions is given by

$$P_r = V^2/4R_r = \frac{(4{\cdot}53)^2 \times 10^{-6}}{4 \times 75} = 0{\cdot}684 \ \text{mW}$$

Hence $P_T/P_r = I^2 . R_r . G/P_r = \dfrac{\frac{1}{4} \times 50 \times 100}{0\cdot684 \times 10^{-3}} = 1\cdot83 \times 10^6$

$$\text{Transmission loss} = 10 \log_{10} P_T/P_r$$

$$= 10 \log_{10} (1\cdot83 \times 10^6)$$

or $\qquad\qquad$ Transmission loss $= 62\cdot63\,\text{dB}$

EXAMPLE 5
A square frame aerial used in direction finding has sides 0·8 m long and has 20 turns. The r.f. resistance of the winding is 20 Ω and its inductance 0·5 mH. It is tuned by a variable capacitor to be in resonance with a wave, the electric field of which is given by $250 \sin 6\pi \times 10^5 t\ \mu\text{V/m}$.

Determine the r.m.s. value of the p.d. across the capacitor when the direction of travel of the wave makes an angle of 30° with the plane of the frame. Explain your solution.

Explain how a frame aerial and a vertical aerial are used together to produce a cardioid diagram.

(L.U.B.Sc(Eng) Tels. Pt. 3, 1960)

Solution
The r.m.s. voltage V induced in the frame aerial is given by

$$V = \frac{2\pi EAN \cos \theta}{\lambda}$$

where E is the r.m.s. electric field intensity at the loop, A is the coil area, N is the number of turns and θ is the angle which the plane of the loop makes with the wavefront. Hence

$$V = \frac{2\pi . 250 . 10^{-6} . (0\cdot8)^2 . 20 \cos 30°}{\sqrt{2} . 10^3}$$

or $\qquad V = 12\cdot3\ \mu\text{V}$

Since the coil is tuned, at resonance, the voltage V_c across the capacitor is

$$V_c = Q . V = \frac{\omega L}{R} . V$$

$$= \frac{6\pi . 10^5 . (0\cdot5) . 10^{-3} . (12\cdot3) . 10^{-6}}{20}$$

or $\qquad V_c = 0\cdot58\ \text{mV}$

The horizontal polar diagram of a vertical monopole is a circle, while the horizontal polar diagram of a frame aerial is a figure of eight where the forward radiation is 180° out of phase with the back radiation. If a frame antenna is placed behind a vertical antenna, the combined field pattern is given by simple pattern addition i.e. the individual patterns are added with their proper signs. This is shown in Fig. 7 where it will be observed that the back radiation is very largely cancelled out and the forward radiation increased, to produce a resultant pattern which is a cardioid.

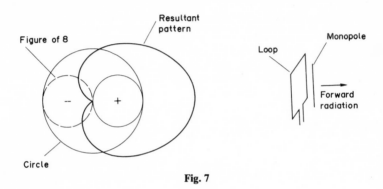

Fig. 7

3
Vertical and horizontal antennas

The simplest form of vertical antenna is the monopole, which is grounded at its lower end. The effect of the earth is to reflect energy incident on it and for a perfectly conducting surface, the reflected energy is considered to come from an image antenna, which replaces the ground. The ground reflection is associated with a reflection coefficient $\rho\underline{/\varphi}$ whose value depends on the ground condition.

Furthermore, for a monopole whose length is greater than $\lambda/8$, the distribution of current on the antenna must be considered. Such antennas are called *resonant* antennas and they support a standing wave. Typical examples are shown in Fig. 8.

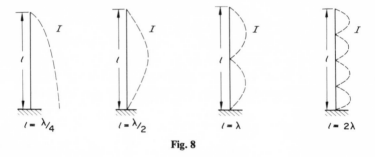

Fig. 8

3.1 Vertical monopole

A grounded $\lambda/2$ monopole is an efficient radiator and its polar pattern can be determined by replacing the ground by the image antenna shown in Fig. 9. For a perfectly conducting earth $\rho\underline{/\varphi} = 1\underline{/0}$ and the phase of the current in the image is the same as in the monopole.

Radiation to a distant point P is received by the direct wave and the ground reflected wave which is considered to come from the image antenna. For an element distant x from the surface, the path difference is $2x \cos \theta$ giving a phase difference φ where

$$\varphi/2\pi = \frac{2x \cos \theta}{\lambda}$$

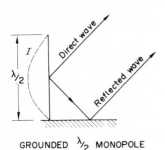

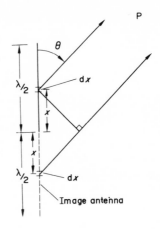

GROUNDED $\lambda/2$ MONOPOLE

Fig. 9

or
$$\varphi = \frac{4\pi x \cos \theta}{\lambda}$$

For the sinusoidal current distribution shown in Fig. 9, the current I at x is given by $I = I_m \sin(2\pi x/\lambda)$ since $I = 0$ when $x = 0$ or $\lambda/2$ and I_m is the maximum value of the *r.m.s* current. Considering the doublet element dx at x, radiation reaching P at a distance r, due to this is

$$dE = \frac{60\pi I \, dx \sin \theta}{\lambda \, . \, r}$$

$$= \frac{60\pi I_m}{\lambda \, . \, r} \sin\left(\frac{2\pi x}{\lambda}\right) . \, dx \sin \theta$$

The resultant field dE_R due to element dx and its image is given by

$$dE_R = 2 \, dE \cos \varphi/2$$

or
$$dE_R = \frac{2 \times 60\pi I_m}{\lambda \, . \, r} \sin\left(\frac{2\pi x}{\lambda}\right) . \, dx \, . \sin \theta \, . \cos\left(\frac{2\pi x \cos \theta}{\lambda}\right)$$

Hence
$$dE_R = E_0 \, . \sin\left(\frac{2\pi x}{\lambda}\right) . \cos\left(\frac{2\pi x \cos \theta}{\lambda}\right) dx$$

where
$$E_0 = \frac{120\pi I_m \sin \theta}{\lambda \, . \, r}$$

Substituting $\alpha = 2\pi/\lambda$ and $\beta = 2\pi \cos \theta/\lambda$ yields

$$dE_R = E_0 . \sin \alpha x . \cos \beta x \, dx$$

or $$E_R = E_0 \int_0^{\lambda/2} \sin \alpha x . \cos \beta x \, dx$$

This is a standard integral* which on integration yields

$$E_R = \frac{60 I_m}{r} \left[\frac{\cos(\pi \cos \theta) + 1}{\sin \theta} \right]$$

and it is plotted in Fig. 10, together with the pattern for $l = \lambda$.

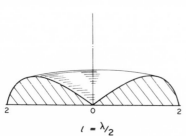
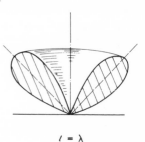

$l = \lambda/2$ $l = \lambda$

Fig. 10

3.2 Horizontal wire (in free space)

A horizontal wire remote from ground and terminated at one end correctly, supports a travelling wave and is called a *non-resonant* antenna. Such an antenna has a uniform current distribution but a progressive phase lag of β radians per unit length. This is illustrated in Fig. 11.

The total radiation reaching a distant point P can be considered as due to little doublets placed end to end along the wire. As the doublets are horizontal, the angle θ defined in Section 2.3 is now measured from the horizontal. For two doublets spaced a distance x apart and of length dx, there is a path difference $x \cos \theta$ and a phase difference $\varphi = (2\pi x \cos \theta)/\lambda$. But the current in the wire is lagging by β radians per unit length where $\beta = 2\pi/\lambda$ and so radiation from the doublet further down the wire at distance x, is lagging that at $x = 0$. Hence, the net phase *lag* $d\varphi$ is

---

* See Geary, A., Lowry, H. V., Hayden, H. A., *Advanced Mathematics for Technical Students. Part 1*, Ch. 3, Longmans Green (1952).

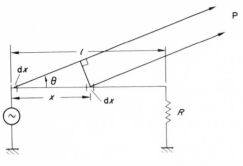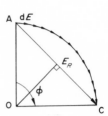

Fig. 11

given by

$$d\varphi = \frac{2\pi x}{\lambda} - \frac{2\pi x \cos\theta}{\lambda} = \frac{2\pi x}{\lambda}(1 - \cos\theta)$$

and the total phase lag between the two ends of the wire for which $x = l$ is given by

$$\varphi = \frac{2\pi l}{\lambda}(1 - \cos\theta)$$

The field produced by each doublet at P is

$$dE = \frac{60\pi I \, dl \sin\theta}{\lambda \cdot r}$$

and the resultant field E_R is obtained by adding the doublet fields end to end with their proper phases.

Hence, from Fig. 11 we obtain

$$\frac{\text{chord AC}}{\text{arc AC}} = \frac{E_R}{\displaystyle\int_0^l dE} = \frac{E_R}{\dfrac{60\pi Il \sin\theta}{\lambda \cdot r}}$$

or

$$\frac{2 \, \text{OA} \sin\varphi/2}{\text{OA} \cdot \varphi} = \frac{E_R \cdot \lambda \cdot r}{60\pi Il \sin\theta}$$

Hence

$$E_R = \frac{120\pi Il \sin\theta \sin[(\pi l/\lambda)(1 - \cos\theta)]}{\lambda \cdot r \cdot (2\pi l/\lambda)(1 - \cos\theta)}$$

or

$$E_R = \frac{60I \sin[(\pi l/\lambda)(1 - \cos\theta)]\sin\theta}{r(1 - \cos\theta)}$$

Comment
If $l = 4\lambda$ we have

$$E_R = 0 \quad \text{when} \quad \sin \theta = 0$$

or
$$\theta = 0, \pi, 2\pi, \ldots$$

and
$$\sin \frac{\pi l}{\lambda} (1 - \cos \theta) = 0$$

with
$$4\pi(1 - \cos \theta) = 0, \pi, \ldots$$

or
$$\theta = 0, 41{\cdot}4°, 60°, 75{\cdot}5°, 90°$$

The field pattern is plotted in Fig. 12 and shows two main forward lobes directed at a low angle.

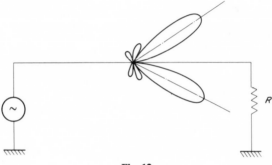

Fig. 12

3.3   Rhombic antenna[4]

The results of the previous section can be used in the design of a horizontal rhombic antenna, which consists of four lengths of horizontal wire laid out in the form of a rhombus. Each wire or *leg* is equal to 4λ at least and the tilt angle φ is then mainly responsible for the resultant field pattern. This is illustrated in Fig. 13.

Each leg radiates in two main lobes which are out of phase by 180° since the field above and below a current conductor is in opposite phase. The lobes a, b, c, d are arranged to be either horizontal or nearly so by the proper choice of tilt angle φ.

In order that the radiation from lobes a and b or c and d may reinforce one another at a distant point in the horizontal plane, the phase difference

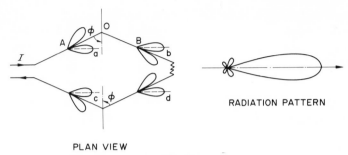

RADIATION PATTERN

PLAN VIEW

Fig. 13

of current along the wire from centre point A to centre point B must be such that it is greater than the *direct* path from A to B by $n\pi$ where n is an odd integer. Hence

$$(\beta . \text{AOB}) - (\omega/c . \text{AB}) = \pi \quad (\text{for } n = 1)$$

where β is the phase-change coefficient along the conductor and ω/c is the phase-change coefficient in free space.

It is found that the value of tilt angle φ for lobes a, b, c, d to be horizontal is slightly greater than that which gives a maximum field at P, but the value of φ is not critical if the leg length equals 4λ or more.

The rhombic antenna therefore gives a highly directional pattern in the horizontal plane and as the forward lobes are directed at a low angle, they are useful for long distance communication. Furthermore, since the tilt angle φ is not critical for leg lengths greater than 4λ, the antenna is a broad-band antenna and can be used over a 2 to 1 frequency range without requiring adjustments.

However, a disadvantage is the loss of half the power in the terminating resistance and at low wavelengths, it requires a large ground area for construction.

EXAMPLE 6
A receiving aerial consists of a horizontal wire 100 m long. It is correctly terminated to prevent reflection at the ends A and B, the termination at B being the feeder to the receiver.

Estimate the ratio of voltages fed to the receiver from horizontally polarised waves arriving at (a) 30° and (b) 60° with the direction AB. The frequency of both signals is 15 MHz. Neglect the effect of the earth.

Describe a rhombic aerial and discuss its merits.

(L.U.B.Sc(Eng) Tels Pt. 3, 1961)

Solution

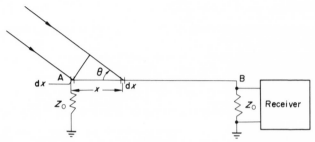

Fig. 14

Let the field strength at the antenna be E_0. The voltage induced per unit length in the antenna is $E_0 \sin \theta$, where θ is the angle which the incident wave makes with the wire.

For two doublets of length dx spaced a distance x apart, the net phase lag is $(\beta x - (2\pi x/\lambda) \cdot \cos \theta) = (2\pi x/\lambda)(1 - \cos \theta)$. As $x = l$ over the wire length, the total phase lag φ is given by

$$\varphi = (2\pi l/\lambda)(1 - \cos \theta)$$

The resultant voltage V_R at the receiver is given by the chord AC where

$$\frac{\text{chord AC}}{\text{arc AC}} = \frac{2 \cdot \text{OA} \sin \varphi/2}{\text{OA} \cdot \varphi} = \frac{\sin \varphi/2}{\varphi/2} \quad \text{(Fig. 11)}$$

where

$$\text{arc AC} = \int_0^l E_0 \sin \theta \cdot \mathrm{d}x = E_0 \sin \theta \cdot l$$

Hence

$$V_R = E_0 \sin \theta \cdot l \cdot \frac{\sin \varphi/2}{\varphi/2}$$

or

$$V_R = E_0(\lambda/\pi)\sin \theta \times \frac{\sin[(\pi l/\lambda)(1 - \cos \theta)]}{(1 - \cos \theta)}$$

Now

$$l = 100 \text{ m}$$

$$\lambda = c/f = \frac{3 \times 10^8}{15 \times 10^6} = 20 \text{ m}$$

and $$l/\lambda = 100/20 = 5$$

(a) For $\theta = 30°$

$$\sin 30° = 0.5$$

$$1 - \cos 30° = 1 - 0.8660 = 0.134$$

$$(\pi l/\lambda)(1 - \cos 30°) = 5\pi \times 0.134 = 120.6°$$

(b) For $\theta = 60°$

$$\sin 60° = 0.8660$$

$$1 - \cos 60° = 1 - 0.5 = 0.5$$

$$(\pi l/\lambda)(1 - \cos 60°) = 5\pi \times 0.5 = 450°$$

If V_1 and V_2 are the *corresponding* voltages at the receiver then we have

$$V_1 = \frac{E_0\lambda \sin 30°}{\pi} \times \frac{\sin 120.6°}{0.134}$$

$$V_2 = \frac{E_0\lambda \sin 60°}{\pi} \times \frac{\sin 450°}{0.5}$$

Hence

$$V_1/V_2 = \frac{\sin 30°}{\sin 60°} \times \frac{\sin(59.4°)0.5}{\sin(90°)0.134}$$

$$= \frac{0.25 \times 0.8607}{0.866 \times 0.134}$$

or $$V_1/V_2 = 1.85$$

The answer to the last part of the question will be found in Section 3.3.

3.4 Horizontal wire (near ground)

For a horizontal wire near ground, the effect of the ground is to cause radiation to be radiated upwards. It requires the use of an image antenna to replace the ground, which is assumed to be perfectly conducting.[5] This is shown in Fig. 15 for a general case of an antenna at a height of $n\lambda$ above a perfectly conducting surface, for which the image has a phase reversal of 180°

For radiation reaching a distant point P, there is a path difference of $2n\lambda \cos \theta$ which gives a phase difference of $2\pi/\lambda \times 2n\lambda \cos \theta$ or $4n\pi \cos \theta$. The total phase lag φ is therefore $(\pi + 4n\pi \cos \theta)$. If the field due to the

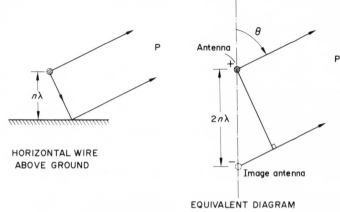

HORIZONTAL WIRE
ABOVE GROUND

EQUIVALENT DIAGRAM

Fig. 15

wire is E_0, the resultant field E_R at P is

$$E_R = 2E_0 \cos \varphi/2$$

$$= 2E_0 \cos\left[\frac{\pi + 4n\pi \cos \theta}{2}\right]$$

or $\qquad E_R = 2E_0 \sin(2n\pi \cos \theta)$

Comments
1. $E_R = 0$ when
   $2n\pi \cos \theta = 0, \pi, 2\pi, --- $ giving
   $\theta = \cos^{-1} 0, \cos^{-1} 1/2n, --- $
2. E_R is a maximum when
   $2n\pi \cos \theta = \pi/2, 3\pi/2, --- $ giving
   $\theta = \cos^{-1} 1/4n, \cos^{-1} 3/4n, --- $
   The resultant field E_R has been plotted in Fig. 16 for heights of $\lambda/2$
and λ. They reveal that the main lobes are now directed upwards and
their number increase as the height above ground increases.

Height function

The height above ground of a vertical or horizontal receiving antenna, has
an effect on the received signal from a distant transmitter, due to interference between the direct ray and the ground reflected ray. This is
illustrated in Fig. 17.

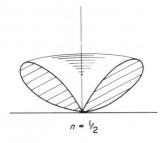
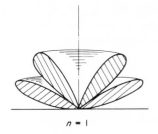

$n = \frac{1}{2}$ $n = 1$

Fig. 16

(a) Vertical case

If the angle of the received beam is θ there is a path difference of $2h_r \sin \theta$ between the received rays which gives a phase difference of $(2\pi \cdot 2h_r \sin \theta)/\lambda$ or $(4\pi h_r \sin \theta)/\lambda$.

Generally, $h_T \gg h_r$ and so $\sin \theta \simeq h_T/d$. Normally also, there is no phase reversal, on reflection from the ground for vertically polarised rays, but at high frequencies in the VHF band and beyond, there is a phase difference of about 180° at low angles of elevation θ.

Hence, the total phase difference between the two rays is $180° \pm (4\pi h_r h_T)/\lambda d$. If the field strength at the transmitter is E_0 then the field strength of either ray at the receiver is E_0/d and the resultant E_R is

$$E_R = 2(E_0/d)\cos\left[\frac{180 \pm (4\pi h_r h_T)/\lambda d}{2}\right]$$

$$= 2(E_0/d)\sin(2\pi h_r h_T)/\lambda d$$

or $$\frac{E_R}{E_0/d} = 2\sin[(2\pi h_r h_T)/\lambda d] \simeq (4\pi h_r h_T)/\lambda d$$

for small angles.

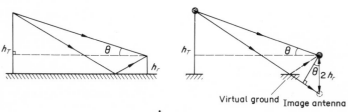

h_T θ h_r

h_T θ θ $2h_r$

Virtual ground   Image antenna

Fig. 17

The ratio $E_R/(E_0/d)$ varies sinusoidally with a maximum value of 2. It depends upon the value of h_r for a given h_T and d. Hence, it represents the height function of the receiving antenna and is plotted in Fig. 18(a).

(b) Horizontal case

Here the path difference is also $2h_r \sin \theta$ giving a phase difference of $(4\pi h_r \sin \theta)/\lambda$. Since there is always a 180° phase reversal for horizontally polarised waves due to the image antenna, the total phase difference is $180 \pm (4\pi h_r \sin \theta)/\lambda$. Hence

$$E_R = 2(E_0/d)\cos\left[\frac{180 \pm (4\pi h_r \sin \theta)/\lambda}{2}\right]$$

or

$$\frac{E_R}{E_0/d} = 2 \sin\left(\frac{2\pi h_r \sin \theta}{\lambda}\right) \simeq (4\pi h_r \sin \theta)/\lambda$$

The expression is similar to that for the vertical case except that the angle of approach θ of the received signal is used instead of the height of the transmitting antenna. The height function $E_R/(E_0/d)$ is therefore the same and is plotted in Fig. 18(a).

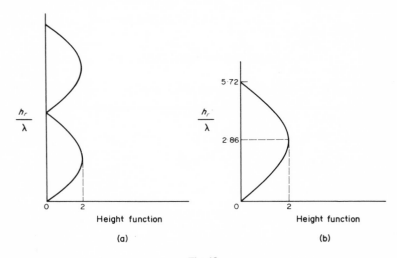

Fig. 18

EXAMPLE 7

Discuss the reasons that cause the strength received by a VHF aerial to vary with height above ground.

Derive an expression for the height function of an aerial erected over plane earth having perfect conductivity.

A horizontally polarised dipole aerial mounted above perfectly conducting plane earth receives a signal wave arriving at an elevation angle of 5°. Plot the height function in wavelengths over a range of heights sufficient to include the first maximum and the first minimum responses.

(C. & G. Adv. Comm. Radio 1969)

Solution

The received signal varies with height above ground because of the interference caused between the direct ray received and the ground reflected ray. As their path lengths are different, depending upon the receiving antenna height, at certain heights the rays reinforce one another and at others they directly cancel one another. Hence, the received signal will vary from zero to a maximum value depending upon h_r, the receiving antenna height.

PROBLEM

The expression for the resultant field strength has been derived in Section 3.4 and is given by

$$E_R = 2(E_0/d)\sin(2\pi h_r \sin \theta/\lambda)$$

or

$$\frac{E_R}{E_0/d} = 2 \sin(2\pi h_r \sin \theta/\lambda)$$

The minimum value of zero occurs when

$$\sin(2\pi h_r \sin \theta/\lambda) = 0$$

In general, $\sin \theta \neq 0$ and so

$$h_r/\lambda = 0$$

or

$$h_r = 0$$

The maximum value of $E_R/(E_0/d)$ is 2 when

$$\sin\left(\frac{2\pi h_r \sin \theta}{\lambda}\right) = 1$$

or
$$\frac{2\pi h_r \sin \theta}{\lambda} = \pi/2$$

or
$$h_r/\lambda = 1/4 \sin \theta = \frac{1}{4 \times 0.0872} = 2.86$$

The height function is plotted in Fig. 18(b).

EXAMPLE 8

Discuss the advantages and disadvantages of travelling wave aerials.

An array consists of two identical rhombic aerials, spaced 100 m apart and in line with a distant transmitter. The aerials are connected to individual loss-free coaxial cables whose outputs are connected in series at the receiver. If the difference in length of the two cables is 100 m, determine the total voltage at the receiver when the transmitter produces a down coming wave at an angle of 50° to the horizontal and induces a voltage of 1 mV at 12 MHz in each aerial. Assume the relative permittivity of the dielectric of the coaxial cables to be unity.

(L.U.B.Sc(Eng) Tels. Pt 3, 1962)

Solution

and support a travelling wave.

Advantages
1. They produce forward radiation only and so no loss of power in the backward direction.
2. They are broad-banded and can be used over a 2 to 1 frequency range without adjustment.
3. They produce low angle lobes which are suitable for long distance communication.

Disadvantages
1. They occupy a lot of space for erection.
2. Half the power is wasted in the terminating resistance approximately.

PROBLEM

The phase lag of the nearest antenna to the receiver due to path difference $d \cos \theta$ is $(2\pi d \cos \theta)/\lambda$. But the phase lag due to the longer cable from the endmost antenna is βd or $2\pi d/\lambda$, since $\beta \simeq 2\pi/\lambda$.

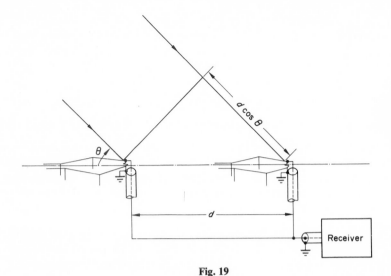

Fig. 19

Hence, the net phase lag between the voltages V_0 induced in the antennas is given by

$$\varphi = 2\pi d/\lambda - (2\pi d \cos \theta/\lambda) = (2\pi d/\lambda)(1 - \cos \theta)$$

The resultant voltage V_R at the receiver is

$$V_R = 2V_0 \cos \varphi/2 = 2V_0 \cos[(\pi d/\lambda)(1 - \cos \theta)]$$
$$= 2 \times 10^{-3} \times \cos\left[\frac{\pi 100}{25}(1 - \cos 50°)\right]$$
$$= 2 \times 10^{-3} \cos 257°$$
$$= 2 \times 10^{-3} \times 0{\cdot}225$$

or $\qquad V_R = 0{\cdot}45 \text{ mV}$

3.5 Half-wave dipole

A very well known and efficient radiator is the half-wave antenna, consisting of two $\lambda/4$ lengths and centre fed. Such a radiator has an input impedance of about 73 Ω and a sinusoidal current distribution. At VHF frequencies and beyond, the dimensions are of convenient length and are popular TV antennas.

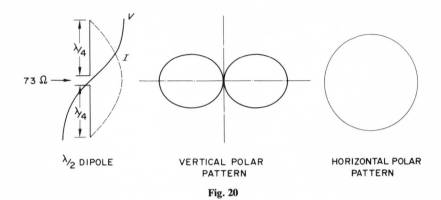

λ/2 DIPOLE   VERTICAL POLAR   HORIZONTAL POLAR
                            PATTERN   PATTERN

Fig. 20

The expression for the vertical polar pattern is derived by considering the antenna as composed of a number of small dipoles placed end to end and is given in Appendix D. Plots of the vertical and horizontal patterns are shown in Fig. 20. For horizontal dipoles, the patterns are similar but interchanged.

EXAMPLE 9
Draw a diagram showing the approximate current distribution along a half-wave dipole which is centre-fed with a current of 2 A (r.m.s.).

The magnitude of the electric field strength (E) at a distance (r) metres from a radiating current element $(I\,dl)$ ampere-metres in the plane perpendicular to and bisecting the length of the element is given by

$$E = \frac{60\pi I\,dl}{\lambda r}\ \text{volts/m}$$

Use this relationship to calculate the electric field at a point in this plane at a distance 10 km from the half-wave dipole.

If the radiation resistance of the dipole is 74 Ω, calculate the gain in dB of the dipole, compared with that of an aerial which radiates uniformly the same total power.

(L.U.B.Sc(Eng) Tels. Pt 3, 1964)

Solution
The expression for E can be applied to a dipole by substituting the value of the effective length l_e of a dipole where $l_e = 2/\pi \times \lambda/2 = \lambda/\pi$. Hence,

the expression for E becomes

$$E = \frac{60\pi I\lambda}{\lambda r\pi} = \frac{60I}{r}$$

where I is the r.m.s. antenna current. Hence

$$E_{\text{rms}} = \frac{60 \times 2}{10 \times 10^3} = 12 \text{ mV/m}$$

The power density at the distant point due to an isotrope radiating uniformly a power $P = I^2 R_r$ is

$$P_d = P/4\pi r^2 = \frac{I^2 R_r}{4\pi r^2} = \frac{4 \times 74}{4\pi \times (10^4)^2}$$

The power density at the same point due to the dipole radiating a total power P is

$$P'_d = E^2/120\pi$$

Hence   Power gain $G = P'_d/P_d = \dfrac{(12 \times 10^{-3})^2}{120\pi} \times \dfrac{(4\pi \times 10^8)}{4 \times 74}$

$$= \frac{144 \times 10^{-6} \times 4\pi \times 10^8}{120\pi \times 296}$$

or $G = 1\cdot62 = 2\cdot1 \text{ dB}$

4
Antenna arrays

The directivity of a vertical antenna in the vertical plane depends upon
its height which can be altered by changing its height above ground.
However, to obtain directivity in the horizontal plane, two or more
antennas must be used to form an array. The resultant field pattern can be
obtained by considering the antennas as point source radiators. It is
assumed that the power fed to a single antenna is the same as that fed to
the array, in order that a fair comparison of gain may be made.

4.1 Two point sources

Consider two point sources spaced a distance d apart in the horizontal
plane and energised by currents in the same phase. At a distant point P
the field strengths of each are approximately equal, but there is a phase
difference due to a path difference as shown in Fig. 21.

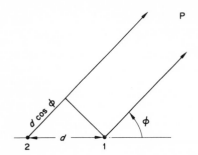
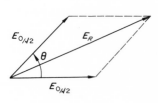

Fig. 21

The difference in path length is $d \cos \varphi$ which produces a phase
difference between the fields E_0 of each source. Hence

$$\frac{\theta}{2\pi} = \frac{d \cos \varphi}{\lambda}$$

or
$$\theta = \frac{2\pi d \cos \varphi}{\lambda}$$

If the field strength due to a single source is E_0 when energised by a power P, then the field strength of each element in an array of two is $E_0/\sqrt{2}$ as each element is energised by a power $P/2$ i.e. the field strength of each element is reduced by $\sqrt{2}$.

The resultant field E_R of the array is given by the parallelogram law where

$$E_R^2 = \frac{E_0^2}{2} + \frac{E_0^2}{2} + 2\frac{E_0^2}{2}\cos\theta = E_0^2(1 + \cos\theta)$$

$$= 2E_0^2\cos^2\theta/2$$

or $$E_R = \sqrt{2}\,E_0\cos\theta/2 = \sqrt{2}\,E_0\cos\left(\frac{\pi d\cos\varphi}{\lambda}\right)$$

In general, if the currents lag or lead each other by angle α then

$$E_R = \sqrt{2}\,E_0\cos\left[\frac{\theta \pm \alpha}{2}\right]$$

the +ve sign being used for a current lag and the −ve sign for a current lead.

4.2 *n* radiators

To generalise the array system, consider a system of n radiators where n will be initially assumed to be 4.

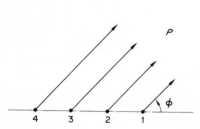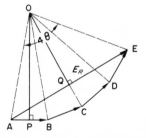

Fig. 22

The radiations from the point sources give a resultant E_R which is the closing side of a polygon. Each side of the polygon is the field due to a single point source in the array, energised by $P/4$ i.e. $E_0/\sqrt{4}$ where E_0 is the field strength of a single point source *alone*, energised by a power P.

Since the phase difference between each element is θ, $\angle AOE = 4\theta$. Now OQ and OP are perpendicular to AE and AB respectively hence

$$\angle AOQ = 2\theta$$
$$\angle AOP = \theta/2$$

and from Fig. 22 we obtain

$$E_R = AE = 2AQ = 2OA \sin(4\,\theta/2)$$

Also $\qquad AP/OA = \sin \theta/2$

or $\qquad OA = AP/\sin(\theta/2) = \dfrac{E_0/2\sqrt{4}}{\sin \theta/2}$

Substituting for OA in E_R yields

$$E_R = \frac{2(E_0/2\sqrt{4})\sin(4\,\theta/2)}{\sin \theta/2} = \frac{E_0}{\sqrt{4}}\left[\frac{\sin(4\,\theta/2)}{\sin \theta/2}\right]$$

In the case of n antennas where n is any integer number we have

$$E_R = \frac{E_0}{\sqrt{n}}\left[\frac{\sin(n\,\theta/2)}{\sin \theta/2}\right] = \sqrt{n}\,E_0\left\{\frac{\sin(n\,\theta/2)}{n \sin \theta/2}\right\}$$

or $\qquad E_R = E_0\left[\dfrac{\sqrt{n}\sin(n\,\theta/2)}{n \sin \theta/2}\right]$

Comments
1. The quantity in the *curly* brackets has a maximum value of 1.
2. The last quantity in the *square* brackets is called the array factor.
3. The maximum value of the array factor is $\sqrt{n}$ for n antennas i.e. the field strength of the array has been increased $\sqrt{n}$ times or its power gain G has been increased by $(\sqrt{n})^2$ or n times.

For an additional phase angle α in the currents, the total phase angle is $\theta \pm \alpha$, the +ve sign being used for lagging currents and the $-$ve sign for leading currents. Hence

$$E_R = E_0\left[\frac{\sqrt{n}\sin\{n(\theta \pm \alpha)/2\}}{n \sin\{(\theta \pm \alpha)/2\}}\right]$$

Typical plots of field strength E_R for various values of n and α are shown in Fig. 23.

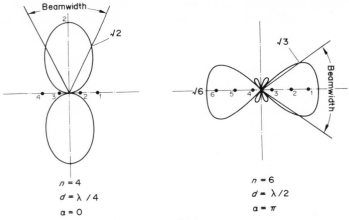

Fig. 23

The pattern obtained with $\alpha = 0$ has its main lobes at right angles to the line joining the point sources and is called a *broadside* pattern, while the pattern obtained with $\alpha = \pi$ has main lobes along the line of centres of the point sources. It is called an *end-fire* pattern. The beamwidth of the patterns is defined as the angle between the 3 dB points and is also shown in Fig. 23. Generally, broadside patterns give narrower lobes than end-fire patterns and are used more often.

EXAMPLE 10

Two half-wave dipoles are mounted with their centres at a height of 10λ above a perfectly conducting ground plane. The dipoles are parallel and normal to the line joining their centres, the distance between centres being 600λ. The first dipole is connected to a matched generator delivering 1 W at a wavelength λ and the second dipole is matched to a receiver. Calculate the power delivered to the receiver for (a) horizontal dipoles and (b) vertical dipoles. The power gain of a dipole in free space is 2·17 dB relative to an isotropic source. Mention any assumptions made.

Explain, with the aid of suitable sketches, the way in which the received power varies in each case with changes in the distance between the dipoles. Indicate on these sketches how the results would be modified if the ground were perfectly conducting. (C.E.I. Comm. Eng. Part 2, Spring 1970)

Solution

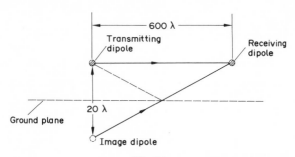

Fig. 24

Since the antennas are 10λ above ground, the effects of mutual impedances between the antennas and ground may be neglected. Considering the antenna as a point source, power reaches the receiving antenna by the direct ray and the ground reflected ray. The latter is assumed to come from the image antenna. The resultant field strength at the receiving dipole is found by the phasor sum of the two fields for the two cases.

Horizontal dipoles

The path difference between the two rays is $600 \cdot 3\lambda - 600\lambda = 0 \cdot 3\lambda$ giving a phase difference of $2\pi/\lambda \times (0 \cdot 3\lambda) = 0 \cdot 6\pi = 108°$. As the image antenna is 180° out of phase with the real antenna, the total phase difference is $\varphi = 180° \pm 108°$. Hence

$$E_R = 2 E_0 \cos\left(\frac{180° \pm 108°}{2}\right) = 2 E_0 \sin 54°$$

where E_0 is the field strength at the receiving antenna due to a single ray (direct or reflected) from the transmitting antenna. Now

$$P_d = E_0^2/120\pi = P/4\pi d^2$$

where P is the transmitted power and P_d is the power density at a point distant d from the transmitter. Hence

$$E_0 = \frac{\sqrt{30GP}}{d}$$

where $G = 1.64$ is the power gain of the transmitting dipole, or

$$E_0 = \frac{\sqrt{30 \times 1.64}}{600\lambda} = \frac{7.014}{600\lambda}$$

with $$E_R = \frac{(2 \times 7.014)}{600\lambda} \times 0.809 = \frac{5.66}{300\lambda}$$

If V is the voltage received by the receiving antenna and the power delivered to the receiving antenna is P_r then $V = E_R \times \lambda/2 = 5.66/600$ and

$$P_r = V^2/(4 \times 75) = V^2/300 = \frac{(5.66)^2}{300 \times (600)^2}$$

or $$P_r = 0.296 \times 10^{-6} \text{ watts}$$

Vertical dipoles

If the angle of elevation of the ground reflected ray is θ then

$$\tan \theta = 10\lambda/300\lambda = \tfrac{1}{30}$$

or $$\theta \simeq 2°$$

For such small angles of elevation, there is also a phase reversal on reflection for vertical dipoles and is accounted for by an image antenna $180°$ out of phase with the real antenna. Hence, the power received will be the same as for a horizontal dipole.

At very large distances from the transmitter, the path difference between the two ray paths is negligible but their phase difference is $180°$. Hence, the power received will be zero for both vertical and horizontal dipoles. As the distance decreases, the power received will increase to a maximum when the path difference between the two rays is equal to $\lambda/2$. At closer distances, the power received will fluctuate about the free space value as in Fig. 49.

The polar patterns of radiation are shown in Fig. 25 as solid lines for a perfectly conducting plane and vertical dipoles. For an imperfectly conducting plane, losses in the ground will soon attenuate radiation

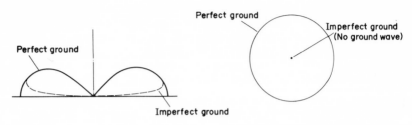

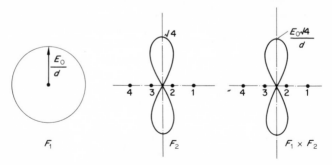

VERTICAL PLANE PATTERN   HORIZONTAL PLANE PATTERN

Fig. 25

received near the ground and is shown in dotted lines in Fig. 25. Horizontal dipoles will not be affected so much by ground effects.

4.3 Pattern multiplication

The expression for the resultant field pattern of an array of point sources can be expressed in two parts, namely

$$E_R = F_1 \times F_2$$

where F_1 is the field pattern of a single point source radiator and F_2 is the array factor for the n radiators. This resultant expression enables the field pattern to be determined for arrays in which the elements may be other than point sources e.g. dipoles, dipoles with reflectors etc. All that is required is the field pattern F_1 of the single radiator and the array factor F_2 of the array with each radiator treated as a point source. The final field pattern in a plane is obtained by multiplying the two field patterns for that plane and is known as pattern multiplication.

Fig. 26

This is illustrated in Fig. 26 for four dipole radiators spaced $\lambda/2$ apart in a horizontal line with co-phased currents. The field pattern of a single dipole in the horizontal plane is a circle, while the array factor comprises two broadside lobes. The resultant field pattern is similar in shape to F_2 with a maximum value of $\sqrt{4}\, E_0/d$.

4.4 Typical arrays

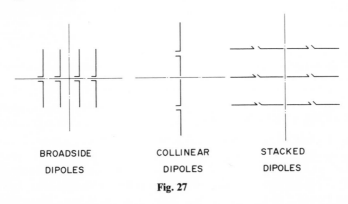

BROADSIDE DIPOLES   COLLINEAR DIPOLES   STACKED DIPOLES

Fig. 27

(a) Broadside dipoles

The co-phased dipoles are placed along a horizontal line, d metres apart. The beamwidth of the broadside lobes is given approximately by

$$\text{Beamwidth} \simeq \frac{\lambda}{nd}\ \text{rads.}$$

and the power gain G is n relative to a single *dipole*.

(b) Collinear dipoles

The co-phased dipoles are stacked vertically end to end with centres approximately d metres apart. The broadside beamwidth is given by

$$\text{Beamwidth} \simeq \frac{\lambda}{nd}\ \text{rads.}$$

and the power gain G is n relative to a single *dipole*.

(c) Stacked dipoles

An $m \times n$ array consists of m co-phased dipoles in a row stacked in n tiers. They can be treated as an array of $m \times n$ point sources giving a gain G of $m \times n$ relative to a single *dipole*. The field pattern in the horizontal plane is obtained by multiplying the field pattern of a single dipole, with that of the array pattern. The array pattern is itself the product of the field pattern due to a horizontal row of point sources times the field pattern of a vertical collinear array of point sources, all the patterns being obtained for the same horizontal plane.

(d) Yagi array[6,7]

If one dipole only of an end-fire array is energised, it may be used in turn to energise an adjacent element of the array. The latter is called a parasitic element and if placed behind the dipole it is known as a *reflector* and if placed in front of the dipoles, it is called a *director*.

At resonance, the dipole length is slightly less than $\lambda/2$ while the reflector is made equal to $\lambda/2$ and the director, slightly less than $\lambda/2$. A reflector placed behind a dipole has a positive reactance and a director placed in front of the dipole, has a negative reactance. Since they are reactive, they have no losses but merely re-radiate energy falling on them to give increased forward directivity in the horizontal plane, as shown in Fig. 28.

The driven antenna and parasitic element behave as coupled circuits with self and mutual inductance, depending upon their lengths and spacing. If V_1 is the voltage applied to the dipole and I_1, I_2 are the

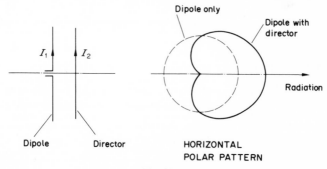

Fig. 28

Table 1

Element spacing (λ)	Z_{12} (ohms)	θ_{12} (degrees)
0·1	70	15
0·15	60	0
0·2	55	−20
0·25	50	−35

respective currents of dipole and parasitic element, then the z parameter equations of the system are

$$V_1 = z_{11}I_1 + z_{12}I_2$$
$$0 = z_{21}I_1 + z_{22}I_2$$

where z_{11}, z_{22} are the self inductances and $z_{12} = z_{21}$ is the mutual impedance between the antenna elements. Hence

$$I_2 = \frac{-z_{12}I_1}{z_{22}} = \left| \frac{z_{12}}{z_{22}} \right| I_1 \, \underline{/180 + \theta_{12} - \theta_{22}}$$

where θ_{12} is the phase angle of z_{12} and depends on the position of the parasitic element and θ_{22} is the phase angle of z_{22} which depends on its length. Hence, by adjusting the position and length of the parasitic element, the current I_2 can either lag or lead I_1. Values of $z_{12}\underline{/\theta_{12}}$ and $z_{22}\underline{/\theta_{22}}$ are given in Tables 1 and 2 respectively.

Table 2

Element length (λ)	Z_{22} (ohms)	θ_{22} (degrees)
0·53	95	+40
0·5	73	0
0·47	74	−10
0·45	84	−30

N.B. The values for Z_{22} are for thin elements.

By using several parasitic elements in an end-fire array, a highly directional pattern can be obtained. It is called the Yagi-Uda antenna and is a popular television and satellite tracking antenna. Usually, only one reflector is necessary but several directors may be employed. The reflector is about 0.5λ long and spaced about 0.15λ to 0.25λ behind the driven dipole whose resonant length is made about 0.46λ long. Directors are usually 0.43λ to 0.4λ long and placed about 0.1λ apart. A gain of 10 dB is attained using 5 or 6 elements which may be increased to 13 dB with about 9 or 10 elements.

As the presence of the parasitic elements tends to load the driven element, its input impedance is reduced. However, by using a folded dipole which has two parallel current paths, the current in each path is halved and so the impedance is increased four times in order to give the same radiated power I^2R as the dipole. Hence, its radiation or input resistance is $4^2 \times 75 = 300\,\Omega$. A typical Yagi arrangement is shown in Fig. 29.

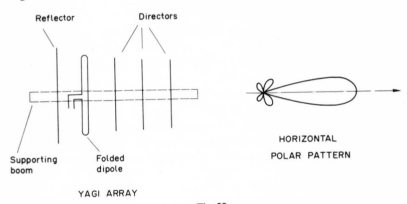

Fig. 29

(e) Slot array[8]

A rectangular slot in an infinite conducting sheet will radiate if fed with energy at the opposite ends of the slot, across its narrow dimension. It behaves like a dipole of the same size as the slot except that the E and H fields are interchanged i.e. a vertical slot radiates a horizontally polarised wave. This is illustrated in Fig. 30 where the E lines are in opposite directions on the two sides of the plane.

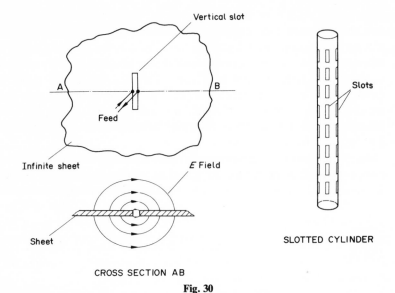

CROSS SECTION AB

Fig. 30

The radiation from one side of the slot may be suppressed by using a finite sheet bent around as a cylinder. By placing four slots symmetrically around the cylinder an omnidirectional pattern is obtained in the horizontal plane. To obtain greater power gain in the horizontal plane, several slots are placed along a vertical line to form a collinear array.

A typical array uses 32 slots placed in tiers of eight and facing N, S, E, W. The eight slots in a tier form a collinear, cophased array of dipoles giving a power gain of 8 × 1·64 relative to an isotropic radiator. Slot lengths are about 0·7λ and spaced λ apart on a cylinder of about 0·6λ diameter. The slots are fed by a common coaxial cable, whose lengths are arranged to ensure cophasing of the slots.

The arrangement is shown in Fig. 30 and such antennas are used for broadcasting VHF or television. The input impedance to the slot is about 470 Ω and the antenna is broad-banded with a ± 4% bandwidth. The beamwidth is determined from the expression

$$\text{Beamwidth} \simeq \frac{\lambda}{nd} = \frac{\lambda}{8\lambda} = 1/8 \text{ rad.} = 7°$$

EXAMPLE 11
Explain the difference between travelling wave and standing wave aerials
and refer to an example of each type.
Eight co-planar vertical dipoles with their centres at half-wavelength
spacings along a horizontal line, carry currents which are equal in
magnitude and phase. Derive an expression for the polar diagram in a
horizontal plane in terms of the field strength E, in this plane due to a
single dipole. Hence, calculate the angle between the two directions of
zero radiation that include the main beam.
What would be the main defect of such an array when used for point-to
point operation? Explain how this could be remedied.

(L.U.B.Sc(Eng) Tels. Pt. 3, 1966

Solution
A *travelling wave* antenna supports a travelling wave pattern as it is
correctly terminated. The voltage is fairly uniform along the whole line.
An example is the rhombic antenna.
A *standing wave* antenna however is not terminated and so a standing
wave pattern exists on the antenna, giving points of maximum and
minimum voltage. An example is the half-wavelength dipole.
The expression for the polar diagram of a cophased vertical array is
given in Section 4.2 by

$$E_R = \frac{\sqrt{n}\, E_0 \sin(n\, \theta/2)}{n \sin \theta/2}$$

where $\theta = (2\pi d \cos \varphi)/\lambda$, d being the spacing between the antennas.
Here, $n = 8$ and $d = \lambda/2$ hence we have

$$\theta = \frac{(2\pi\lambda/2)\cos \varphi}{\lambda} = \pi \cos \varphi$$

with
$$E_R = E_0\sqrt{8}\left[\frac{\sin(4\pi \cos \varphi)}{8 \sin(\pi \cos \varphi/2)}\right]$$

Now $E_R = 0$ when

$$\sin(4\pi \cos \varphi) = 0 \quad \text{if} \quad \sin(\pi \cos \varphi/2) \neq 0$$

or $4\pi \cos \varphi = 0, \pi, 2\pi, \ldots$

and $\cos \varphi = 0, \tfrac{1}{4}, \tfrac{1}{2}, \ldots$

Since cos $\varphi = 0$ is not allowed as the denominator will tend to zero, the first minimum occurs when

$$\cos \varphi = \tfrac{1}{4} = 0\cdot25$$

or $$\varphi = 75\cdot5°$$

and $$\text{Beamwidth} = 2 \times 14\cdot5 = 29°$$

The main defect for point-to-point communication is the loss of power in the radiation from the back lobe. This could be remedied by using reflectors behind the dipoles or better still, a metal reflecting screen. The extra gain in the latter case is about 6 dB.

EXAMPLE 12
Explain what is meant by the principle of pattern multiplication used in determining the polar diagram of an aerial array.

An aerial array consists of four identical isotropic point sources of equal amplitude and phase, located at the corners of a square having a 3 $\lambda/4$ diagonal length. Making use of the principle of pattern multiplication, determine the polar diagram of the array in the plane containing the sources. Plot the normalised polar diagram.

(C & G, Adv. Comm. Radio 1968)

Solution
The answer to the first part is given in Section 4.3.

PROBLEM
The overall polar diagram is given by pattern multiplication with $E_R = F_1 \times F_2$. The array factor F_1 is that of a single point source radiator and is a circle of radius E_0, where E_0 is the field strength of the single radiator. This is shown in Fig. 31(a). The array factor F_2 is obtained

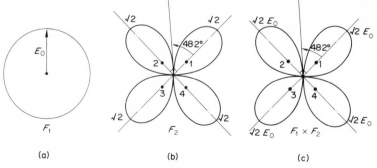

(a) (b) (c)

Fig. 31

as follows:

The resultant field pattern of sources 1 and 3 is

$$E'_R = \sqrt{2} \cos \theta/2$$

where $\theta/2 = \pi d \cos \varphi/\lambda = \pi \times 3\,\lambda/4 \times \cos \varphi/\lambda = 3\pi \cos \varphi/4$

or $$E'_R = \sqrt{2} \cos[3\pi \cos \varphi/4]$$

Now E'_R is a maximum when

$$\cos(3\pi \cos \varphi/4) = 1$$

or $$3\pi \cos \varphi/4 = 0$$

Hence $$\cos \varphi = 0$$

or $$\varphi = \pi/2$$

relative to the *centre line* joining sources 1 and 3.

Now E'_R is a minimum when

$$\cos(3\pi \cos \varphi/4) = 0$$

or $$3\pi \cos \varphi/4 = \pi/2$$

Hence $$\cos \varphi = \tfrac{2}{3} = 0.666$$

or $$\varphi = 48.2°$$

The field pattern for point sources 2 and 4 is identical to this relative to the *centre line* joining sources 2 and 4. Hence, the combined field pattern for the four isotropic sources gives the array factor F_2 which is plotted in Fig. 31(b). The overall polar pattern for the four point sources is given by

$$E_R = F_1 \times F_2$$

which is plotted in Fig. 31(c).

5
Microwave antennas

At microwave frequencies, antenna dimensions which are convenient to handle are several wavelengths in size and it is therefore possible to apply geometrical optics[9] in the analysis of such antennas. Hence, many of the results obtained by geometrical optics (such as reflection and refraction effects), can be applied to microwave antennas with considerable advantage and lead to fairly accurate solutions of antenna problems.

The main application of microwave antennas for the frequency range of 1 GHz and beyond is in point-to-point communication and radar. In such systems, high gain antennas with very narrow beamwidths in one or more planes are required and may be achieved with antennas of reasonable size. The most important microwave antennas are horns, parabolic reflectors, slot radiators and to a lesser extent, dipoles, dielectric rods and lenses.

Of primary importance are the field patterns produced by these antennas. Pattern measurements may be made in the near-field *Fresnel* region or in the far-field *Fraunhofer* region. Most results are required for the Fraunhofer region for which the distance r is $> 2D^2/\lambda$ where D is the largest dimension of the antenna. In the Fresnel region, for which r is $< 2D^2/\lambda$, interference effects give results depending upon the distance r and so are normally avoided in antenna measurements.

Since many microwave antennas possess an aperture e.g. horns and parabolic reflectors, it is usually necessary to use the concept of effective aperture A which is directly related to antenna gain G as was shown in Section 1.6. From geometrical optics it can be shown that the larger the aperture, the greater is the power gain and narrower is the beamwidth.* This is essential in applications such as radar tracking or in satellite communications.

5.1 Horn antennas[10,11]

An open-ended waveguide is an inefficient radiator of energy because of the impedance mismatch at the mouth. However, it can be improved by

---

* See Section 5.1.

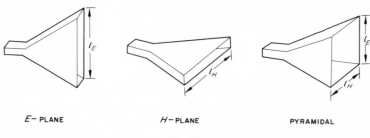

E– PLANE   H–PLANE   PYRAMIDAL

Fig. 32

flaring out the sides into a horn, thereby producing an impedance match between the waveguide and that of the intrinsic impedance of free space. Horns flared in the E or H plane are called sectoral horns and if flared in both E and H planes, are called pyramidal horns, as shown in Fig. 32.

The horn mouth is essentially a rectangular aperture and its radiation pattern is determined by the amplitude and phase distribution over the aperture. For a plane uniform amplitude distribution, the pattern may be derived using Huygens principle[12] and it has the well-known $\sin u/u$ distribution in the far-field region as shown in Fig. 33, where $u = \pi a \sin \theta / \lambda$.

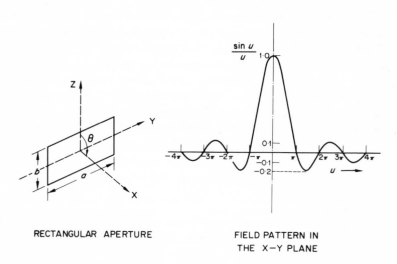

RECTANGULAR APERTURE   FIELD PATTERN IN
                               THE X–Y PLANE

Fig. 33

For a horn however, the wavefront near the horn is spherical, but at a great distance it is nearly plane wave over a small area. E-plane horns give slightly narrower beamwidths than H-plane horns, because the E vector is uniform over the aperture length l_E whereas it varies sinusoidally along length l_H for a H-plane horn. Both however, show a phase distribution across the aperture, because the path length along the axis is less than that along the flared edge. This leads to a loss of gain and a modified polar pattern.[13]

The effect may be reduced by using a long, gradual flare, at the expense of increasing horn size and weight. In practice, a compromise is necessary since the larger the aperture, the larger is the directivity, but the phase distribution across the aperture is greater. Hence, there are optimum design conditions for any particular horn which are given in the shape of design curves. Typical practical data for horns will be found in Southworth.[13] Of special interest, is the power gain G and 3 dB beamwidth of a pyramidal horn and is given by

$$G = 7 \cdot 5A/\lambda^2$$
$$B.W. = \lambda/d \text{ rads.}$$

where λ is the wavelength, A is the area of the horn aperture and d is the width of the aperture in the plane of the beamwidth. Furthermore, important features of horns are their narrow beamwidth capabilities, the possibility of obtaining broad-band characteristics by using ridged structures, a high side lobe level and higher order moding in the flared section.

5.2 Parabolic reflectors[14]

The dimensions of the radiating surfaces of many parabolic antennas are large compared to the wavelength used and so, geometrical optics which is based upon rays and wavefronts, may be used to study certain aspects of these antennas. Some important properties of the reflector can be studied using ray optics, while others require the use of electromagnetic field theory.

(a) Geometrical aspects

A parabolic surface has the useful property of being able to convert a diverging spherical wavefront into a parallel plane wavefront, thereby producing a highly focussed or narrow pencil beam. The property of focussing which is usually akin to light rays and lenses, can be equally well applied to parabolic reflectors at microwave frequencies. The parabolic surface satisfies the equation

$$y^2 = 4fx$$

where y is any ordinate at a point on the surface, f is the focal length of the surface and x is the corresponding abscissa. This is illustrated in Fig. 34.

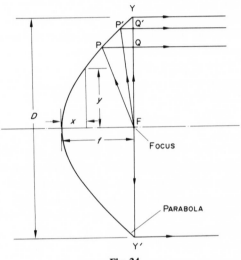

Fig. 34

In Fig. 34, all the rays leaving F after reflection from the surface emerge parallel to one another and reach the plane YY' at the same time and so form an *equiphase* wavefront. Hence, we have $FP + PQ = FP' + P'Q' =$ a constant. Usually the focus F is at the plane wavefront. Hence, $D = 4f$ is the diameter of the aperture. More generally, the f/D ratio is important and in practice varies from 0·25 to 0·5. If this ratio is too low, the reflector is not properly illuminated by radiation leaving the focal point and if too large, radiation is lost over the rim of the parabolic surface as *spill-over*. Of great importance are the reflector and feed system. The overall far-field radiation pattern depends upon various factors which must be considered to give the best results.

(b) Radiation pattern

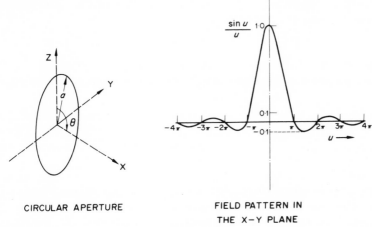

CIRCULAR APERTURE

FIELD PATTERN IN
THE X-Y PLANE

Fig. 35

The parabolic reflector is essentially a circular aperture illuminated by radiation with a certain amplitude and phase distribution. In the simple case of uniform amplitude and phase distribution, the result yields a pattern with a $2J_1(u)/u$ distribution as derived in Weeks[15] which has lower sidelobe levels than the rectangular aperture. This is shown in Fig. 35, where $u = \pi a \sin \theta / \lambda$.

In the practical design of reflectors a more exact solution[16] to the problem is obtained by considering the overall field pattern, the required aperture illumination which will produce this, the primary feed pattern which together with the reflector gives the required aperture illumination and finally, mechanical considerations, tolerances etc. The problem is tackled by first establishing the primary feed pattern as a functional relation. This field induces surface currents in the reflector which are determined and the field produced by these surface currents at the distant point is integrated over the whole surface, to give the overall secondary far-field pattern required.

(c) Feeds

Various feeds used with reflectors are shown in Fig. 36 and the design of a suitable feed system for the reflector constitutes an important task. The

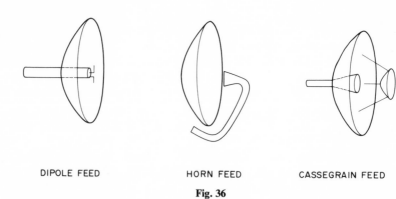

DIPOLE FEED   HORN FEED   CASSEGRAIN FEED

Fig. 36

purpose of the feed is to illuminate the reflector over its entire surface and to avoid spill-over radiation. Moreover, the illumination must be such so as to give the radiation pattern required. Various techniques may be employed and one of the usual is to use an illumination taper i.e. a falling-off of primary beam intensity towards the edges of the reflector as a cosine taper or linear taper. An optimum taper exists therefore which gives best gain, side-lobe level, noise temperature etc.

Furthermore, positioning the feed is another important consideration. The exact phase centre may best be determined experimentally for say, a horn feed. In other cases, blockage may occur viz. in the Cassegrain feed due to the subreflector and supports causing loss of gain, increased side-lobe levels etc. This may be overcome by using the offset or hoghorn type of reflector.

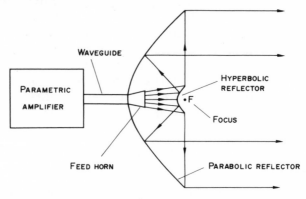

Fig. 37

However, the Cassegrain reflector[17,18] shown in Fig. 37 is the most popular type used in satellite communications. It consists of a sub-reflector which is hyperbolic in shape and illuminated by a horn. This construction enables the low-noise amplifier to be mounted close to the feed and leads to a low noise temperature.

(d) Construction

Reflectors are usually large in size involving complex mechanical construction for rigidity and flexibility. The usual size is around 80 to 90 feet in diameter to give the required figure of merit G/T, where G is the antenna gain and T is its noise temperature. The reflector is usually constructed of panels and the surface finish determines the maximum usable frequency e.g. surface accuracy must be about $\lambda/8$ to $\lambda/16$. Deviations from the parabolic surface lead to a loss of gain and was studied by Ruze,[19] who treated random variations as producing a desired component (specular reflection) and a scattered (diffuse) component. The diffuse component reduces gain and increases side-lobe level and the analysis involves a study of the r.m.s. deviations of the surface and upon the correlation interval of the irregularities. Further consideration must also be given to deformation produced by wind, snow and sheer weight of surface. This also leads to loss of gain and increased side lobe levels.

5.3 Slot antennas[20]

Slots cut in the side wall of a waveguide may or may not radiate energy depending upon their position. Slots cut in the broad wall of a waveguide along its length are used for VSWR measurements and are not intended to radiate. However, slots when cut in the narrow wall of a waveguide, will radiate if they intercept the paths of wall current flow. This is shown in Fig. 38.

A horizontal slot $\lambda/2$ in length behaves as a $\lambda/2$ vertical dipole and will radiate a vertically polarised wave as was mentioned in Section 4.4(e). By combining a number of such slots suitably spaced by $\lambda_g/2$, a broadside array is produced. The spacing is $\lambda_g/2$ because the wall currents reverse every $\lambda_g/2$ along the waveguide wall and so produce the required co-phasing if the slots are tilted in opposite directions.

The amount of slot radiation and its phasing depend upon the angle which the slot makes with the vertical and also on its position along the side

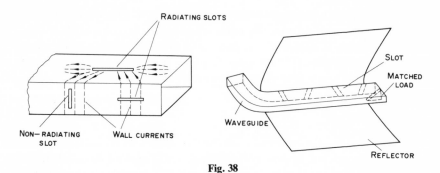

Fig. 38

wall. By combining these various aspects, the side wall radiation can be adjusted to give the required beam pattern for an antenna. Radiating slots serve as useful feeds for the parabolic cylindrical reflector shown in Fig. 38, which is intended to give a narrow horizontal lobe, but a much broader vertical lobe, since in marine radar applications, it must take into account the vertical movement of the ship.

5.4 Lenses[21]

The two main types of microwave lenses are the dielectric lens and the metal-plate lens. They are able to convert a spherical wavefront into a

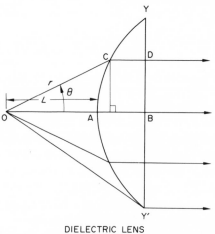

DIELECTRIC LENS

Fig. 39

plane wavefront and so provide greater directivity. This is illustrated in Fig. 39 and 40.

(a) Dielectric lens

Its action is similar to that of an optical lens. The different paths through the lens require the same time of travel and so an equiphase wavefront is produced across YY'. If v is the velocity of the wave in the lens and c is the velocity of light, the condition for an equiphase wavefront is given by

$$\frac{OA}{c} + \frac{AB}{v} = \frac{OC}{c} + \frac{CD}{v}$$

or

$$\frac{OA}{c} + \frac{AB - CD}{v} = \frac{OC}{c}$$

If L is the focal length of the lens and $n = c/v$ is the refractive index of the dielectric we obtain

$$\frac{L}{c} + \frac{(r \cos \theta - L)}{v} = r/c$$

or

$$\frac{L}{c} + \frac{n(r \cos \theta - L)}{c} = r/c$$

Hence

$$r = \frac{(n - 1)L}{(n \cos \theta - 1)}$$

This equation gives the shape of the lens surface facing the source of radiation 0. As $n > 1$ for a dielectric, the surface is a hyperbola.

(b) Metal-plate lens

Its action depends upon the fact that the phase velocity of a wave travelling between parallel plates is greater than that through free space. If the parallel plates are spaced a distance b apart and are of infinite length along the a dimension, they form a waveguide propagating an H_{01} mode with $\lambda_c = 2b$. Hence

$$\frac{1}{\lambda^2} = \frac{1}{\lambda_c^2} + \frac{1}{\lambda_g^2}$$

$$= \frac{1}{(2b)^2} + \frac{1}{\lambda_g^2}$$

or
$$\frac{1}{\lambda_g} = \left[\frac{1}{\lambda^2} - \frac{1}{(2b)^2} \right]^{1/2}$$

The phase velocity v_p of the wave between the plates is related to the guide wavelength λ_g by

$$v_p/c = \lambda_g/\lambda$$

or
$$v_p = \frac{\lambda_g c}{\lambda}$$

If n is the refractive index we obtain

$$n = c/v_p = \lambda/\lambda_g = \sqrt{1 - (\lambda/2b)^2}$$

Hence, $n < 1$ and the surface of the lens which is given exactly by the same expression as for the dielectric lens becomes

$$r = \frac{(1 - n)L}{1 - n \cos \theta}$$

and so the surface facing the source 0 is elliptical in shape. As the expression for n is dependent on λ it is frequency sensitive and a technique used to reduce this effect is known as *zoning*. The zoned plates used are progressively reduced in thickness by steps as shown in Fig. 40.

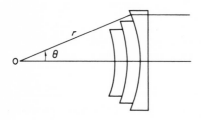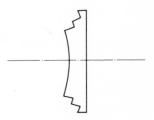

METAL–PLATE LENS   ZONED PLATE

Fig. 40

5.5   Antenna measurements

The main antenna measurements are those concerning its impedance, power gain, polar diagrams and noise temperature.

(a) Impedance

If the impedance presented by an antenna is Z_a when connected to a transmission line of characteristic impedance Z_0, a VSWR is set up on the line, which can be determined using standard techniques. The position of a minimum on the line is then noted and the shift of this minimum, when the antenna is shorted out. From the Smith chart,* the value of Z_a can be determined quite easily.

(b) Power gain

A standard horn whose gain is known may be used in this measurement. The power received from a transmitting source is first measured with the standard horn at the receiver and then the power received by the test horn placed at the receiver. The ratio of the powers received gives the gain of the test horn relative to an isotropic antenna.

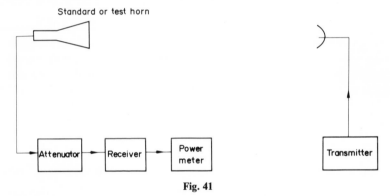

Fig. 41

(c) Radiation pattern

This is the most important characteristic of an antenna. A convenient method is to transmit from a fixed distant secondary antenna ($r > 2D^2/\lambda$) and to rotate the primary antenna whose pattern is to be determined. The field strength and angular position of the primary antenna are noted at various angles.

To reduce reflections from nearby objects, a highly directional transmitting antenna may be used, such as a parabolic reflector rather than a horn.

* See *Wave Transmission* p. 35 by the same author.

This is especially important when measuring low side lobe levels of the primary antenna. Absorbing material may be placed at appropriate points to reduce object reflections.

The pattern of the primary antenna is determined for various polarisations of the incident radiation and so the secondary antenna should have variable polarisation.

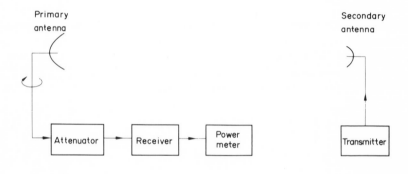

Fig. 42

(d) Noise temperature

The noise power received by an antenna is associated with its *effective* noise temperature T_e and is equal to kT_eB where B is the relevant receiver bandwidth and k is Boltzmann's constant. T_e may be determined by comparing it with a known standard noise source whose noise temperature T_s is known.

The standard noise source is first connected via a low noise receiver to a power meter and the noise power N_s is noted. The antenna is then connected to the low-noise receiver and the noise power N_e is noted on the power meter. From the ratios of the two noise powers N_s and N_e respectively, T_e may be determined from the relationship $N_s/N_e = Y$ with $Y = (T_s + T_R)/(T_e + T_R)$, where T_R is the noise temperature of the receiver. Hence

$$T_e = [T_s + T_R(1 - Y)]/Y$$

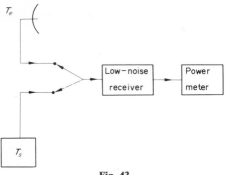

Fig. 43

EXAMPLE 13
The power gain G and effective receiving area A, of an aerial are related
by the equation

$$G = \frac{4\pi A}{\lambda^2}$$

where λ is the operating wavelength.

Explain what is meant by the properties G and A of an aerial and
indicate the steps required to derive the equation.

A satellite communication system operates at a wavelength of 6 cm
with a spherical balloon of equivalent echoing area 10 m² and identical
aerials of 50 dB gain for transmission and reception. The transmitter
power is 10 kW, the receiving aerial noise temperature is 20 K and the
receiver noise temperature is 30 K. The distances from the balloon to the
transmitting and receiving aerials are 2000 and 3000 km respectively.

Calculate the signal-to-noise ratio at the receiver output assuming a
receiver bandwidth of 10 kHz. ($k = 1.38 \times 10^{-23}$ J/K).

(C.E.I. Part 2, Comm. Eng, Spring 1968)

Solution
The answer to the first part of the question will be found in the text in
Chapter 1 and Appendix A.

PROBLEM
The field strength at the balloon is given by

$$E_B = \frac{\sqrt{30 P_T G}}{d_1}$$

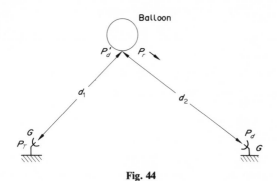

Fig. 44

The power density P_d' at the balloon is

$$P_d' = E_B^2/120\pi = \frac{30P_TG}{d_1^2 \cdot 120\pi} = \frac{P_TG}{4\pi d_1^2}$$

The reflected power P_r from the balloon is

$$P_r = P_d'\sigma = \frac{10P_TG}{4\pi d_1^2}$$

The field strength E_R at the receiver is

$$E_R = \frac{\sqrt{30P_r}}{d_2}$$

The power density P_d at the receiver is

$$P_d = E_R^2/120\pi = \frac{30P_r}{120\pi d_2^2} = \frac{P_r}{4\pi d_2^2} = \frac{10P_TG}{(4\pi d_1 d_2)^2}$$

The received signal is given by

$$S_i = P_dA \quad \text{where} \quad A = G\lambda^2/4\pi$$

or $\quad$ $$S_i = \frac{10P_TG^2\lambda^2}{(4\pi)^3d_1^2d_2^2} = \frac{10 \cdot 10^4 \cdot 10^5 \cdot 10^5 \cdot 36 \cdot 10^{-4}}{4(16) \cdot \pi^3 \cdot 4 \cdot 10^{12} \cdot 9 \cdot 10^{12}}$$

or $\quad$ $S_i = 5{\cdot}05 \times 10^{-17}$ watts

The input noise power $N_i = kT_eB$ where

$$N_i = 1{\cdot}38 \times 10^{-23} \times 20 \times 10^4$$

or $\qquad\qquad N_i = 2.76 \times 10^{-18}$ watts

The effective noise temperature T_e of the receiver is related to the receiver noise figure F by

$$T_e = (F - 1)T_0 \quad \text{where} \quad T_0 = 290 \text{ K}$$

or $\qquad\qquad F = 1 + T_e/T_0 = 1 + \frac{30}{290} = 1.1$

Now $\qquad\qquad F = \dfrac{S_i/N_i}{S_0/N_0}$

Hence $\qquad\qquad S_0/N_0 = \dfrac{5.05 \times 10^{-17}}{2.76 \times 10^{-18}} \times \dfrac{1}{1.1} = 16.5$

or $\qquad\qquad S_0/N_0 = 10 \log_{10}(16.5) = 12.18 \text{ dB}$

6
Electromagnetic waves

Radio communication is due to the propagation and reception of electromagnetic waves[22] in space. Such waves which travel with the velocity of light, possess an electric field and a magnetic field at right angles to one another and to the direction of propagation. The direction of the **E** vector determines polarisation and most radio waves are vertically polarised.

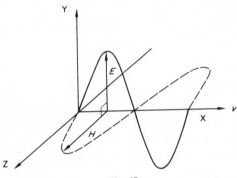

Fig. 45

6.1 Propagation paths

An electromagnetic wave can be launched from an antenna such as a dipole. The antenna can be considered as a two-wire transmission line opened up and the electric and magnetic field lines appear as in Fig. 45. The induction field near the antenna varies inversely as the *distance squared* and quickly dies away, but the radiation field varies inversely as the *distance only* and can be picked up far away from the antenna. As light is an electromagnetic wave, a radio wave is simply treated as a *ray* leaving the antenna like a light ray.

The main types of radio waves are the ground wave, sky wave and space wave. The ground wave consists of the direct wave which travels near the ground from transmitter to receiver and combines with a similar ground reflected wave. At greater distances from the transmitter, the ground reflected wave cannot be received (due to the earth's curvature) and a *surface* wave alone is received as the total ground wave. A sky wave on the other hand, travels upward towards the ionosphere and may be reflected or refracted back from the ionosphere. A space wave is a direct wave which combines with a ground reflected wave, but the point of reception is high up in space such as for an aircraft or satellite.

These waves are affected very much like light waves. The sky wave travelling through the ionosphere is largely refracted and if the refraction is sufficient, the wave is returned back to earth. Alternatively, reflection may occur at a single layer of the ionosphere. On the other hand, radio waves travelling near the earth's surface induce earth currents causing eddy current losses and are quickly attenuated especially at high frequencies. Furthermore, in the upper atmosphere (troposphere), water vapour droplets absorb energy causing attenuation, while in the ionosphere, the ionised particles also absorb energy causing considerable attenuation.

6.2   The ionosphere

The bending of electromagnetic waves down to earth requires the presence of ionised layers in the upper atmosphere. These ionised layers are produced by the ionisation of the rarefied gases present and is due to radiations from the sun. The radiations are mainly ultra-violet rays, γ rays and cosmic particles such as electrons and protons. Due to this, the ionosphere takes on a stratified character called the D, E, F_1 and F_2 layers.

The D-layer is found occasionally at a height of 50 to 100 km in the daytime and is of little importance. The E-layer (Kennelly-Heaviside layer) is a relatively permanent layer at about 100 km. The ionic density is strongest during the day and may almost vanish at night, due to the recombination of ions. The F-layer (Appleton layer) is also more or less permanent at about 300 km. In winter, it divides into the F_1 and F_2 layers and is subject to erratic variations. Apart from seasonal variations, sun-spot activity causes further magnetic storms and consequent radio fadeouts. The layers are found to exist at *virtual heights* which can be measured by radar echoes. Thus, a sky wave which is refracted by the F_2-layer along the path TUR as in Fig. 46 appears to be reflected from a single layer at the point V located at the virtual height h. Figure 46 also illustrates the ionic densities and heights of the various layers.

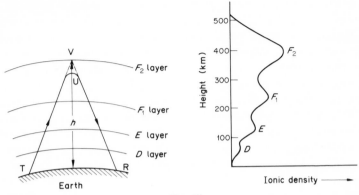

Fig. 46

At vertical incidence, waves upto a certain frequency will be reflected back from the *E*-layer. This is called the lower critical frequency f_c. Above this frequency, they will pass through the *E*-layer and be reflected from the *F*-layer which is denser. Likewise for the *F*-layer there is a critical frequency above which the wave will penetrate the layer and is lost in space. This higher frequency is called the higher critical frequency. The penetration frequency is a measure of the ionisation density because the higher the frequency, the greater must be the density of the ions to be able to reflect back the waves, the relationship is given by

$$f_c = 9\sqrt{N}$$

where N is the number of ions/m³ for the particular layer and f_c is the critical frequency for that layer.

For any two points on the earth; there is a maximum frequency which can be used. A higher frequency would penetrate either the *E* or *F* layers or both. This is called the *maximum usable frequency* f_{muf}. Associated with the two places is the angle of incidence *i* as illustrated in Fig. 47, where the maximum usable frequency is related to the critical frequency f_c for a given layer by the relationship

$$f_{muf} = f_c \sec i = f_c\sqrt{1 + d^2/4h^2}$$

where *d* is the distance between the transmitter and receiver (assuming a flat earth) and *h* is the virtual height of the reflecting layer.

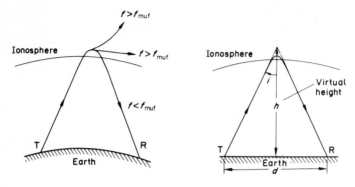

Fig. 47

6.3 Refractive index

The ionosphere can be associated with a refractive index n defined by the relation

$$n = \frac{\text{velocity of the wave in vacuo}}{\text{velocity of the wave in the medium}}$$

At first, consider the ionosphere as an ionised region containing N free electrons per m^3 and let an electromagnetic wave of field strength $E = E_0 \sin \omega t$ be incident on it. As a first approximation, neglect the effect of the earth's magnetic field on the electrons and electron-electron collisions.

The force on the electron is given by

$$F = -eE_0 \sin \omega t$$

or

$$m_0 \frac{d^2x}{dt^2} = -eE_0 \sin \omega t$$

where x is the electron displacement, m_0 is the electron rest mass and $-e$ is the electronic charge.

Integration then yields

$$m_0 \frac{dx}{dt} = eE_0/\omega \cos \omega t$$

where the constant of integration is omitted as only changes in time are of interest.

The electron velocity leads to a conduction current i_c where $i_c = -Ne\,dx/dt$. According to Maxwell's equations, the displacement current i_d must also be considered and is given by

$$i_d = \frac{\partial D}{\partial t} = \frac{\partial}{\partial t}(\varepsilon_0 E) = \frac{\partial}{\partial t}\left[\varepsilon_0 E_0 \sin \omega t\right]$$

or
$$i_d = \omega\varepsilon_0 E_0 \cos \omega t$$

The total current density is $(i_c + i_d)$ where

$$i_c + i_d = \omega E_0 \cos \omega t\left[\varepsilon_0 - \frac{Ne^2}{m_0\omega^2}\right]$$

The expression in brackets amounts to the effective permittivity ε of the medium where

$$\varepsilon = \varepsilon_0 - \frac{Ne^2}{m_0\omega^2}$$

Now the phase velocity in the medium is $1/\sqrt{\mu\varepsilon}$ and in vacuo it is $1/\sqrt{\mu_0\varepsilon_0}$, hence

$$n = \frac{1/\sqrt{\mu_0\varepsilon_0}}{1/\sqrt{\mu\varepsilon}} = \sqrt{\varepsilon/\varepsilon_0}$$

Since $\mu \simeq \mu_0$ for the ionosphere,

Hence
$$n = \sqrt{1 - \frac{Ne^2}{m_0\varepsilon_0\omega^2}}$$

Substituting $e = 1{\cdot}6 \times 10^{-19}$ C, $m_0 = 9 \times 10^{-31}$ kg and $f_c = 9\sqrt{N}$, yields the result

$$n = \sqrt{1 - \frac{81N}{f^2}} = \sqrt{1 - (f_c/f)^2}$$

where f_c is the critical frequency for the layer considered. Since $n < 1$, this means that an incident wave is refracted *away* from the normal in the ionosphere and so eventually bends back to earth.

In Appendix E the effect of collisions and the earth's magnetic field are taken into account.

EXAMPLE 14

A plane electromagnetic wave is propagated at a frequency of 10 MHz into an ionised but otherwise field-free medium. The electron density increases in the direction of propagation. The positive ions may be regarded as effectively immobile. Taking the electronic mass and charge to be 9×10^{-31} kg and $1\cdot6 \times 10^{-19}$ C and the permittivity of free space to be $\frac{1}{36} \times 10^{-9}$ farad/m, determine the electron density at which the phase velocity of the wave tends to infinity and explain what happens physically at this point. Prove any formula used.

(L.U.Ext.B.Sc(Eng) Tels. 1954)

Solution
The refractive index n of the ionosphere was derived in Section 6.3 and is given by

$$n = \sqrt{1 - 81N/f^2}$$

When the phase velocity of the wave in the medium tends to infinity, this corresponds to $n \to 0$. Hence

$$0 = 1 - \frac{81N}{f^2}$$

or

$$N = f^2/81$$

$$= \frac{(10^7)^2}{81}$$

or

$$N = 1\cdot25 \times 10^{12} \text{ ions/m}^3$$

At this point in the ionosphere, the wave folds back on itself and returns to earth.

EXAMPLE 15

What is meant by the *virtual height* and *critical frequency* of an ionospheric layer? If the relative permittivity for a layer is given by

$$\varepsilon_r = 1 - \frac{Ne^2}{\varepsilon_0 m \omega^2}$$

where N is the number of electrons per cubic metre, e and m are the electron charge and mass respectively, develop an expression for the critical frequency and calculate the electron density required to return a signal at 12 MHz incident on the bottom of the layer at an angle of 26° to the normal.

$$(e = 1.59 \times 10^{-19} \text{ C}, m = 9 \times 10^{-31} \text{ kg}, \varepsilon_0 = 8.85 \times 10^{-12} \text{ F/m})$$

(C.E.I. Comm.Eng.Part 2, Spring 1971)

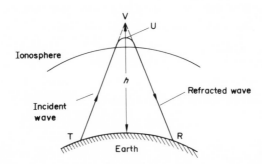

Fig. 48

(a) Virtual height

When a sky wave is incident on an ionospheric layer it is refracted along the path TUR as in Fig. 48. The incident and refracted rays can be extended along the dotted lines to intersect at a point V. Hence, it appears that the sky wave has been reflected from a single ionospheric layer at V, which is defined to be at the virtual height h.

(b) Critical frequency

A wave incident *vertically* on an ionospheric layer is refracted back to earth when the frequency is low. As the frequency is increased, a value is reached when the wave *just* penetrates the layer. This frequency of penetration is called the critical frequency for that layer.

PROBLEM
The critical frequency of a layer occurs when the refractive index n just vanishes. Hence

$$n = \sqrt{\varepsilon_r} = 0$$

with

$$1 - \frac{Ne^2}{\varepsilon_0 m \omega_c^2} = 0$$

or $\qquad f_c^2 = \dfrac{Ne^2}{4\pi^2\varepsilon_0 m}$

$$= \frac{N \times [1 \cdot 59 \times 10^{-19}]^2}{4\pi^2 \times 8 \cdot 85 \times 10^{-12} \times 9 \times 10^{-31} \times 10^3}$$

$$= 81N$$

or $\qquad f_c = 9\sqrt{N}$

At the critical frequency, the refracted ray is just parallel to the earth's surface and $\sin r = 1$. Hence

$$n = \frac{\sin i}{\sin r} = \sqrt{1 - \frac{Ne^2}{\varepsilon_0 m\omega^2}} = \sqrt{1 - \omega_c^2/\omega^2}$$

or $\qquad \sin i = \sqrt{1 - \left(\dfrac{f_c}{f}\right)^2} = \sqrt{1 - \dfrac{81N}{f^2}}$

Hence $\qquad \dfrac{81N}{f^2} = 1 - \sin^2 i = \cos^2 i$

or $\qquad N = \dfrac{f^2 \cos^2 i}{81}$

Substituting $f = 12\ \text{MHz}$ and $i = 26°$ yields

$$N = \frac{(12 \times 10^6)^2 \times (0 \cdot 8988)^2}{81}$$

or $\qquad N = 1 \cdot 44 \times 10^{12}\ \text{electrons/m}^2$

6.4   Characteristics of radio waves

Propagation characteristics of radio waves vary enormously from world-wide transmission to line-of-sight transmission. Physical factors limiting propagation are (a) curvature of the earth (b) resistivity and dielectric constant of the earth and sea (c) tropospheric absorption. In the frequency range from 50 kHz to about 50 MHz radio waves may be grouped as long waves (50 kHz–550 kHz), medium waves (550 kHz–1·5 MHz) and short waves (1·5 MHz–30 MHz).

Long waves

The ground wave is predominant for short distances, but sky wave reception is possible at greater distances and may combine vectorially with the ground wave. Ground wave attenuation is exponential and is given by a semi-empirical formula. It is small at low frequencies. For sky wave propagation, the ionosphere behaves as a flat conducting plate reflecting waves with little attenuation, if angles of incidence are less than the critical value. During the day, the sky wave is largely absorbed in the ionosphere and so night reception is better than day reception.

Long waves are employed largely in ship navigation, direction finding and for some limited broadcast services, since the frequency range is only about 500 kHz. It is therefore restricted to services requiring reliability, freedom from fading and simple geometrical paths which are not seriously affected by the atmosphere.

Medium waves

Ground attenuation increases exponentially due to the semi-empirical formula and so the range is limited to a maximum of about 250 km in daytime. The sky wave is usually absorbed during the day but at night, ranges up to 8000 km are possible for intervals with some fading. Since the ionosphere is highly selective as regards frequency, phase relationships between sidebands continually change, causing distortion to speech and music.

Medium waves are largely used for coverage to limited areas mainly for broadcasting entertainment. Some frequencies are allocated to aircraft and the police, for limited purposes.

Short waves

The ground wave is quickly attenuated at high frequencies and so reception is mainly by sky wave. A definite *skip distance* exists between the end of the ground wave and the return of the sky wave, where no reception is possible. Sky wave propagation depends mainly on ionospheric conditions, maximum usable frequency, critical frequency and the angle of incidence. Moreover, refraction in the ionosphere depends upon n, the refractive index of the ionosphere, which is closely related to ionic density and frequency.

In practice, frequencies are allocated for various purposes and judicious choice of frequency for day and night use is essential. Frequencies for long distance communication lie between 3 MHz and 30 MHz.

If the received signal is the phasor sum of ground and sky wave, when the sky wave fluctuates, the received signal fluctuates and is called fading. In the case of short waves, fading is due mainly to fluctuations in the strength of the received sky wave, caused by ionospheric conditions. Alternatively, *selective* fading may occur when certain sideband frequencies fluctuate only, being attenuated more than others. It may be overcome by single sideband reception or the use of diversity reception.

6.5 VHF propagation[23,24]

(a) Line-of-sight transmission

Frequencies generally greater than 30 MHz penetrate the ionosphere and do not return to earth. Hence, propagation at VHF frequencies and above is mainly by the space wave (direct wave plus ground reflected wave). The direct wave travels through the troposphere and a major influence on propagation is the refractive index of the troposphere.

Since the received signal is the sum of the direct and ground reflected waves, there will be a phase difference which will produce an interference pattern showing maxima and minima as illustrated in Fig. 49.

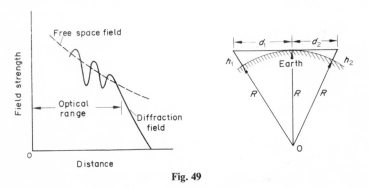

Fig. 49

Neglecting the ground reflected wave and tropospheric refraction as a first approximation, VHF reception is mainly by the direct wave and the maximum distance of reception is determined by the heights of the transmitter and receiver antennas as in Fig. 49.

If R is the earth's radius and h_1 and h_2 the antenna heights, then

$$d_1^2 = (R + h_1)^2 - R^2 = 2h_1R + h_1^2 \simeq 2h_1R$$
$$d_2^2 = (R + h_2)^2 - R^2 = 2h_2R + h_2^2 \simeq 2h_2R$$

Hence, $d = d_1 + d_2 = \sqrt{R}(\sqrt{2h_1} + \sqrt{2h_2})$ metres which is the maximum range.

The effect of tropospheric refraction is to cause the direct ray to travel a curved path between transmitter and receiver, giving a somewhat greater maximum radius. This may be accounted for by using a modified earth radius as shown in the next section.

(b) Tropospheric refraction

The troposphere is the region of the atmosphere adjacent to the earth and extending upwards to about 10 km. It is in this region that clouds are formed and the temperature in the troposphere decreases with height at the rate of $6.5°/km$. Wave propagation in the troposphere beyond the line-of-sight distance can result from several mechanisms. These are diffraction, normal and abnormal refraction and tropospheric scatter.

The direct wave is refracted because the refractive index of the troposphere decreases with height and a consequent downward bending of the radio waves. The refractive index has a value of approximately 1.0003 at the earth's surface. It is usual to use a scaled up value of n called the refractivity defined by $N = (n - 1)10^6 \simeq 300$. The refractivity[25] can be evaluated from the equation

$$N = \frac{77.6}{T}\left[P + \frac{4810p}{T}\right]$$

where P is the total pressure in millibars, p is the partial pressure of water vapour in millibars and T is the absolute temperature in kelvins (K). Typical values at the earth's surface are $P = 1000$ mb, $p = 10$ mb and $T = 300$ K.

Of chief significance as far as bending of radio waves is concerned, is the *gradient* of the refractive index i.e. the rate of change of n with height. If a uniform gradient is assumed, the downward bending may be accounted for by using straight ray paths and an *effective* earth radius greater than the actual radius.

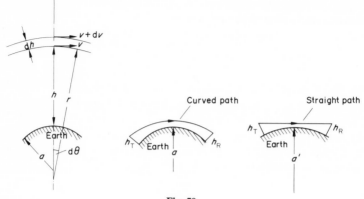

Fig. 50

To obtain the effective radius a', let r be the radius of path at a height h above the earth's surface of radius a and v the wave velocity at that height. From Fig. 50 we obtain

$$r \, \mathrm{d}\theta = v \, \mathrm{d}t$$

$$(r + \mathrm{d}h) \, \mathrm{d}\theta = (v + \mathrm{d}v) \, \mathrm{d}t$$

with
$$\mathrm{d}h \, \mathrm{d}\theta = \mathrm{d}v \, \mathrm{d}t$$

or
$$\mathrm{d}\theta/\mathrm{d}t = \mathrm{d}v/\mathrm{d}h$$

If the refractive index at height h is n, then the velocity of the wave at that height is $v = c/n$ which yields

$$\mathrm{d}v/\mathrm{d}h = (\mathrm{d}v/\mathrm{d}n)(\mathrm{d}n/\mathrm{d}h) = -(c/n^2)(\mathrm{d}n/\mathrm{d}h)$$

$$= -(v/n)(\mathrm{d}n/\mathrm{d}h)$$

As $n \simeq 1$ $\mathrm{d}v/\mathrm{d}h = -v(\mathrm{d}n/\mathrm{d}h)$

Now
$$r = \frac{v}{\mathrm{d}\theta/\mathrm{d}t}$$

and substituting for $\mathrm{d}\theta/\mathrm{d}t$ the expression obtained previously gives,

$$r = -\frac{1}{\mathrm{d}n/\mathrm{d}h}$$

If the radius of curvature of the ray path is defined as ρ then

$$\rho = 1/r = -\mathrm{d}n/\mathrm{d}h$$

The curvature of the earth's surface is $1/a$ and so the relative curvature ρ' of the earth to a straight ray path is

$$\rho' = 1/a - 1/r = 1/a - (-\mathrm{d}n/\mathrm{d}h)$$

or

$$1/a' = 1/a + \mathrm{d}n/\mathrm{d}h$$

where a' is the new effective earth radius as shown in Fig. 50. The value of $\mathrm{d}n/\mathrm{d}h$ under normal conditions is about $-0{\cdot}039 \times 10^{-6}$ per metre while a is taken as 6370 km. Hence, we obtain $a' = 8490$ km or $a' = \frac{4}{3}a$.

However, under *abnormal* conditions, a modified refractive index n' must be used which depends upon the height of the layer h. It is related to the normal refractive index n as shown in Burrows[24] by the expression

$$n' = (n + h/a)$$

(c) Ducting

In conditions when the gradient of the refractive index is not uniform and considerable divergence occurs, superrefraction or trapping is possible and is particularly important in the lower parts of the troposphere. It gives rise to an atmospheric *duct* in which the wave is trapped and travels well beyond the horizon, giving good signals even at great distances from the transmitter. Trapping usually occurs at centimetre wavelengths. This is shown in Fig. 51.

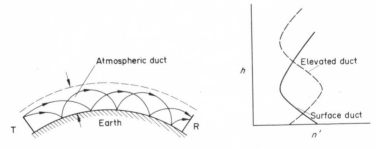

Fig. 51

When refracting conditions cause trapping, then the technique of using an effective earth radius breaks down and a modified refractive index $n' = n + h/a$ must be used. A duct may be at ground level or at an elevated level and is associated with a negative value of $\mathrm{d}n'/\mathrm{d}h$. The negative value is due to a temperature inversion which is caused by hot air rising and cold air taking its place underneath. The phenomenon is observed in hot regions like the Mediterranean and the Carribean, where there are large sea areas.

6.6   Scatter propagation[26,27]

At VHF and UHF frequencies, beyond-the-horizon transmission is possible using scatter mode propagation. The three possible types are (a) tropospheric scatter (b) ionospheric scatter (c) meteor trails.

(a) Tropospheric scatter

Scattering is assumed to be from eddies or *blobs* of air masses in the troposphere, caused by turbulence within the common volume of transmitting and receiving antenna beams. If λ is large compared to the eddies, the scattering is in all directions giving some back scattering, but if λ is small compared to the eddies, mainly forward scattering occurs into a cone at low angles θ as in Fig. 52. More recent theory suggests reflections from discontinuous, inhomogeneous layers, caused by sharp variations in the dielectric constant.

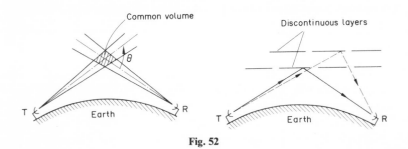

Fig. 52

The received power is considerably less than the free space value being typically about -150 dB at 100 miles for 40 MHz. The received signal also shows a slow variation and a rapid variation in strength because of of the random scattering effect. The frequencies suitable are 400 MHz–5 GHz with a useful range of about 300–600 km, with the maximum range being suitable for narrow band Telephony. Hence, high power transmitters using 1–10 kW are necessary with diversity reception.

(b) Ionospheric scatter

This is due to the lower ionosphere being turbulent in the *E*-layer. It produces *blobs* of ionised regions probably of meteoric origin. Since the scattering centres are much higher than for the previous case, greater range is possible up to 2000 km, the useful range being 1000–2000 km.

Signals fluctuate fairly rapidly but do not show complete fadeout as other conventional ionospheric waves. The frequency range suitable is 30 MHz–60 MHz mainly with transmitters using about 50 kW together with diversity reception.

(c) Meteor trails

Ionisation columns produced by meteors in the *E*-layer region are capable of reflecting waves. The meteor trails are either of low density or high density. The probability of occurrence of the former is high and chiefly accounts for this type of propagation.

The signals received are available at all times using moderate transmitter powers e.g. 1 kW at 30 MHz with a range from 400–2000 km.

In all the above scattering mechanisms, high gain antennas must be used, such as are attainable with parabolic reflectors having diameters between 10–20 metres. Typical antenna gains lie between 30–50 dB.

6.7   Satellite communications[28,29]

The scatter methods described above are subject to frequency and distance limitations. Hence, in more recent years, artificial satellites have also been used extensively, using line-of-sight propagation.

In general, two types of communication satellites have been used (*a*) the *passive* satellite which merely reflects the incident radiation such as the 'Echo' balloons and (*b*) the *active* satellite which is really a repeater in space receiving on one frequency and transmitting on another. Typical examples are Telstar, Syncom and the more recent series of Comsat satellites. The frequencies employed are usually in the UHF and SHF (microwave) bands, so that antennas in the satellites may be of convenient size.

Such communication satellites may be placed in various orbits and at different heights. They are known as synchronous or asynchronous orbits as shown in Fig. 53. Synchronous satellites are more popular since they

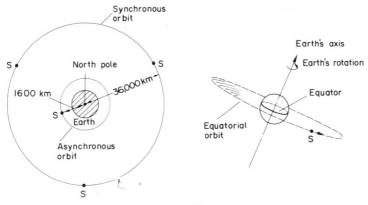

Fig. 53

are placed in a 36 000 km equatorial orbit and they appear stationary to an observer on earth, thus enabling fixed transmitting and receiving antennas to be used. Moreover, 24-hour coverage is possible using only three such satellites in the same synchronous orbit.

Table 3

Satellite	Frequency (GHz)	Total channels	Date
Intelsat 1			
(Early Bird)	4–6	240	1965
Intelsat 2	4–6	240	1967
Intelsat 3	4–6	1200	1969
Intelsat 4	4–6	5000	1971

Path loss however, is between 150–200 dB and so for commercial use, they require large ground-based antennas, usually of about 30 metres diameter. Furthermore, to attain the necessary S/N specifications for the system, low-noise amplifiers such as masers or parametric amplifiers must be used with cryogenic systems. Though such satellites are being exploited more and more, it looks as if communication satellites will continue to be complimentary to other existing systems for many years to come. Details of some communication satellites launched to date are given in Table 3.

Problems

1 Explain what is meant by the effective area (aperture) of an aerial.
   A microwave radio link transmitter has an output power P_T and an aerial gain G_T. The receiving aerial, distant r metres, has a power gain G_R. If the gain of an aerial is $4\pi/\lambda^2$ times its effective aperture, derive an expression for the received power.
   Given that the transmitter power is 8 W and the transmitting and receiving aerials have gains of 30 dB, calculate the received power for a 40 km link operating at a wavelength of 3 cm.
   (L.U.B.Sc(Eng) Tels. Pt. 3, 1968)

2 Explain the basis on which an approximate distribution of the current in a resonant aerial can be made. Deduce an expression for the current at a distance l from the midpoint along a centre-fed dipole in terms of the feed current if the length is half a wavelength. The radiation field at a point P, distant s from a current element is given by

$$E = \left[\frac{60\pi I \; dl \sin \theta}{s\lambda}\right] \cos(\omega t - 2\pi s/\lambda) \text{ volts/metre}$$

where θ is the angle between the element dl and the line to the point, and λ is the wavelength. Form an integral to give the total field at P due to the dipole.
   Evaluate the integral for the value of θ equal to 90° and hence calculate the effective length of the dipole for transmission in this direction.
   (L.U.B.Sc(Eng) Tels. Pt. 3, 1970)

3 Define the term *radiation resistance* of an aerial and explain its practical significance as a property of transmitting and receiving aerials.
   An aerial has a radiation resistance of 70 Ω and is supplied with a current of 4 A. The power transmitted in the direction of maximum radiation is twice that from an omnidirectional aerial radiating the same power. Determine the wave intensity and the electric field strength at a point P, 30 miles from the aerial in this direction.
   An identical aerial is situated at P and directed towards the transmitter. The effective length of each aerial is 2 m. Calculate the

maximum power available to a receiver connected to the second aerial, and the overall transmission loss in dB.

(L.U.B.Sc(Eng) Tels. Pt. 3, 1963)

4 Develop an equation for the range of a radar system in terms of transmitted pulse energy, radar cross-section of the target, carrier wavelength λ and minimum detectable signal. Discuss the significance of such factors as antenna gain, receiver noise figure etc., in determining the range.

A 10 cm radar is operating with a peak output power of 200 kW and a pulse length of 1 μs. Calculate the noise figure of the receiver if an aircraft of radar cross-section 80 m^2 is to be detected at a range of 200 km. The aerial gain G may be assumed to be 30 dB and the effective area A of the aerial is related to G by $A = G\,\lambda^2/4\pi$. It may be assumed that satisfactory detection results when the signal/noise ratio is 6 dB or better. $(kT = 4\cdot11 \times 10^{-21}$ J).

(C.E.I. Pt. 2, Comm. Eng. 1971)

5 Discuss the factors which influence the type of transmitting aerial needed for a radio transmission system.

Three high frequency vertical dipoles are energised in phase and are evenly spaced $\lambda/4$ along a horizontal straight line as an isolated broadside array. Determine from first principles the beamwidth between the half power points of the main lobe of the horizontal polar diagram. (L.U.B.Sc(Eng) Tels. Pt. 3, 1969)

6 Sketch the current distribution in a half-wave dipole. For a Hertzian dipole of length δl excited with a current $I_0 e^{j\omega t}$ the radiation field at range r expressed in polar (spherical) co-ordinates (r, θ, φ) is given by

$$E_\theta = \frac{j60\pi}{\lambda}\frac{\delta l}{r} I_0 \sin\theta e^{j\omega(t - r/c)}$$

Hence obtain the radiation field for a half-wave dipole. Sketch the polar diagram and show that the beam width is 78°. Explain how the polar diagram is modified when the half-wave dipole is used in a Yagi array.

Note. $\displaystyle\int e^{ax}\cos nx\,dx = \frac{e^{ax}}{(a^2 + n^2)}(a\cos nx + n\sin nx)$

(C.E.I. Pt. 2, Comm. Eng. Spring 1971)

7 A six-element receiving aerial array consists of a horizontal line of vertical dipoles equally spaced by 1·5 m, the outputs of which are

added in phase. What will be the horizontal directional pattern of the array at a frequency of 100 MHz? What are the significant properties of this pattern? Sketch the directional behaviour of the array at (a) a frequency of 400 MHz (b) a frequency of 40 MHz.

(C.E.I. Pt. 2, Comm. Eng., Oct. 1968)

8 Explain briefly the meaning of the terms gain, polarisation, bandwidth and beamwidth as applied to an aerial array. An aerial array comprises three colinear $\lambda/2$ dipole elements with a centre spacing of λ. Determine the gain of the array and the beamwidth of the major lobe between 3 dB points in the plane containing the elements.

(C & G Adv. Comm. Radio, May 1969)

9 Explain what is meant by 'the equivalent noise temperature' of an aerial.

Discuss the manner in which account could be taken of this factor in the design of a troposoheric scatter link.

The receiving aerial of a tropospheric scatter link has a gain of 46 dB at a frequency of 1 GHz. Calculate the power flux density of a signal incident at the aerial that would yield a signal-to-noise ratio of 40 dB in a bandwidth of 5 MHz at the aerial terminals.

(C & G Adv. Comm. Radio, June 1971)

10 Derive expressions for the phase and group velocities of an electromagnetic wave propagated through an ionised gas, the effects of collisions being neglected. Describe the significance of the fact that at some frequencies the effective permittivity of the gas is zero.

What relevance have these results to the propagation of radio waves in the ionosphere? How are the conclusions modified if collisions and the earth's magnetic field are taken into account?

(C.E.I. Pt. 2, Comm. Eng., Oct. 1968)

11 Derive an approximate expression for the maximum distance of line-of-sight transmission, for a radio system, in terms of the heights of the transmitting and receiving aerials, if the radius of the earth is assumed to be 6350 km.

If a transmitting aerial is located at the top of a tower 200 m above the surface of the earth, determine the maximum distance at which an aircraft, flying at an altitude of 3000 m, will be able to receive signals from the transmitter. Assume that only (line-of-sight) propagation is involved. If the transmitting aerial has a power gain of 13 dB in the direction of the aircraft and the power radiated is 400 W determine the electric field strength of the signal at the aircraft.

(L.U.B.Sc(Eng) Tels. Pt. 3, 1970)

12 With the aid of diagrams, discuss the influence that meteorological and topographical conditions have on the propagation of UHF waves.

Distinguish between sub-refraction, super-refraction and ducting.

Explain clearly why the fading characteristics over a land path usually differ from those over a sea path.

(C & G Adv. Comm. Radio, June 1971)

Answers

1 0·0284 W
2 λ/π
3 0·0774 W/m^2
   5·4 mV
   0·416 μW
   94·3 dB
4 2·05
5 107°
7 Gain = 6
   B.W. = 18°
8 Gain = 3
   B.W. = 19°
9 0·24 × 10^{-12} W/m^2
11 245·6 km
   2 mV

References

1 SHEPHERD, J., MORTON, A. H. and SPENCE, L. F. *Higher Electrical Engineering.* Pitman (1970).

2 CONNOR, F. R. *Wave Transmission.* Edward Arnold (1972). Chapter 4.

3 FOSTER, D. Loop antennas with uniform current, *Proceedings Institute of Radio Engineers,* 32 (October 1944). Page 603.

4 FOSTER, D. Radiation from rhombic antennas, *Proceedings Institute of Radio Engineers,* (October 1937). Page 1327.

5 KRAUS, J. D. *Antennas.* McGraw-Hill (1950). Chapter 11.

6 YAGI, H. Beam transmission of ultra short waves, *Proceedings Institute of Radio Engineers* (June 1928). Page 715.

7 KING, R. Self and mutual impedances of parallel identical antennas, *Proceedings Institute of Radio Engineers,* 40 (August 1952). Page 981.

8 RIBLET, J. *Proceedings Institute of Radio Engineers,* 35 (1947). Page 474.

9 SILVER, S. *Microwave Antenna Theory and Design.* Radiation Laboratory Series, McGraw-Hill (1949). Volume 12.

10 BARROW, B. L. and CHU, L. J. Theory of the electromagnetic horn. *Proceedings Institute of Radio Engineers,* 27 (1939). Page 51.

11 RHODES, D. R. *Proceedings Institute of Radio Engineers,* 36 (September 1948). Page 1101.

12 PARIS, D. T. and HURD, F. K. *Basic Electromagnetic Theory.* McGraw-Hill (1969).

13 SOUTHWORTH, G. C. *Principles and Applications of Waveguide Transmission.* Section 10.1. D. Van Nostrand (1950).

14 CUTLER, C. C. Parabolic antenna design for microwaves. *Proceedings Institute of Radio Engineers,* 27 (November 1947). Page 1284.

15 WEEKS, W. L. *Antenna Engineering.* McGraw-Hill (1968). Chapter 6.

16 COLIN, R. E. and ZUCKER, F. J. *Antenna Theory.* McGraw-Hill (1969). Volumes 1 and 2.

17 HANNAN, P. W. Microwave antennas derived from the Cassegrain telescope, *Transactions Institute of Radio Engineers,* AP-9 (March 1961). Page 140.

18 WILLIAMS, F. W. *High efficiency antenna reflector.* Microwave Journal. (July 1965). Page 79.

19 RUZE, J. The effect of aperture errors on the antenna radiation pattern. *Nuovo Cimento.* 9, Supplement 3, (1952). Page 364.

20 RADIO RESEARCH LABORATORY STAFF, *Very High Frequency Techniques,* 1 McGraw-Hill (1947). Chapter 7.

21 KOCK, W. E. *Proceedings Institute of Radio Engineers,* 34 (1946). Page 828 and *Bell System Technical Journal,* 27 (1948). Page 58.

22 TERMAN, F. E. *Radio Engineers Handbook.* McGraw-Hill (1943).

23 MATTHEWS, P. A. *Radio Wave Propagation (VHF and Above).* Chapman & Hall (1965).

24 BURROWS, W. G. *VHF Radio Propagation in the Troposphere.* Intertext Books (1968).

25 DAVID, P. and VOGE, J. *Propagation of Waves.* Pergamon Press (1969).

26 BAILEY, P. K., BATEMAN, R. and KIRBY, R. C. Radio transmission at VHF by scattering and other processes in the lower atmosphere. *Proceedings Institute of Radio Engineers*, 43 (1955). Page 1181.

27 RICE, P. and HERBSTREIT, J. W. Tropospheric propagation, *Advances in Electronics and Electron Physics*. 20 (1964). Page 199.

28 MUELLER, G. R. and SPANGLER, E. R. *Communication Satellites*. Wiley and Sons (1964).

29 GATLAND, K. W. *Telecommunicution Satellites*. Iliffe Books (1964).

Appendices

Appendix A : Effective aperture A

Consider a plane wave incident on a doublet having an assumed aperture A. If the power density at the doublet is P_d, the received power P_r is given by

$$P_r = P_d A$$

If V is the induced voltage in the doublet of length dl and radiation resistance R_r, then

$$V = E \, dl$$

where E is the field intensity at the doublet. The power available from the doublet is P_r where

$$P_r = V^2/4 \, R_r = P_d A$$

or

$$V^2/4R_r = (E^2/120\pi)A$$

with

$$A = \frac{120\pi E^2 (dl)^2}{4E^2 R_r} = \frac{30\pi (dl)^2}{R_r}$$

Since

$$R_r = 80\pi^2 \left(\frac{dl}{\lambda}\right)^2$$

Hence

$$A = \frac{30\pi (dl)^2 \lambda^2}{80\pi^2 (dl)^2}$$

or

$$A = \frac{3\lambda^2}{8\pi}$$

Since the received power must increase proportionately as the gain of an antenna increases, for two antennas with gains G_1, G_2 and effective apertures A_1, A_2 we have

$$G_1/A_1 = G_2/A_2$$

Comparing an isotrope for which $G_1 = 1$ with that of a doublet having gain $G_2 = 1 \cdot 5$, we obtain

$$A_1 = \frac{1}{1 \cdot 5} \times \frac{3\lambda^2}{8\pi} = \lambda^2/4\pi$$

for an isotrope.

Hence, in general, for an antenna with gain G *relative* to an isotrope and effective aperture A, we obtain

$$\frac{1}{G} = \frac{\lambda^2/4\pi}{A}$$

or

$$A = G\frac{\lambda^2}{4\pi}$$

Appendix B: Radiation field

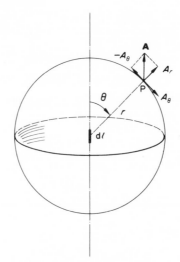

Fig. 54

Let the magnetic vector potential at a point P distant r from the doublet at angle θ be **A**. From field theory we have

$$\mathbf{A} = \int_V \frac{\mu \mathbf{J}(t)\, \mathrm{d}v}{4\pi r}$$

where $\mathbf{J}(t)$ is the current density at time t in the volume $\mathrm{d}v$. For a doublet of length dl carrying a current $I_0 e^{j\omega t}$, we have

$$\int_V \mathbf{J}(t) \, dv \equiv \int_l I_0 e^{j\omega t} \, d\mathbf{l}$$

Hence
$$\mathbf{A} = \frac{\mu I_0 e^{j\omega t} \, d\mathbf{l}}{4\pi r}$$

Since
$$A_\theta = -A \sin \theta$$
$$A_r = A \cos \theta$$

Hence
$$A_\theta = \frac{-\mu I_0 e^{j\omega t} \, dl \sin \theta}{4\pi r}$$

$$A_r = \frac{\mu I_0 e^{j\omega t} \, dl \cos \theta}{4\pi r}$$

These equations do not take into account the time t required for changes at the doublet to travel to P, where $t = r/c$ and c is the velocity of light. Hence, it leads to the *retarded potential* **A** at P where

$$\mathbf{A} = \int_V \frac{\mu \mathbf{J}(t - r/c) \, dv}{4\pi r}$$

and the expressions for A_θ and A_r become

$$A_\theta = \frac{-\mu I_0 e^{j\omega(t - r/c)} \, dl \sin \theta}{4\pi r}$$

$$A_r = \frac{\mu I_0 e^{j\omega(t - r/c)} \, dl \cos \theta}{4\pi r}$$

Now, from field theory we have
$$\text{curl } \mathbf{A} = \mathbf{B}$$
or
$$\text{curl } \mathbf{A}/\mu = \mathbf{H}$$

The magnetic field component is H_φ and for spherical co-ordinates we have

$$H_\varphi = -\frac{1}{r}\left[\frac{\partial}{\partial r}(rA_\theta/\mu) - \frac{\partial}{\partial \theta}(A_r/\mu) \right]$$

or
$$H_\varphi = \frac{1}{r}\left[\frac{1}{4\pi}\{-j \, dl \sin \theta I_0 e^{j\omega(t - r/c)}(-\omega/c)\} + \frac{1}{4\pi r}\{I_0 e^{j\omega(t - r/c)} \, dl \sin \theta\} \right]$$

or $$H_\varphi = \frac{I_0 \, dl \sin \theta}{4\pi} \left[\frac{j\omega e^{j\omega(t-r/c)}}{cr} + \frac{e^{j\omega(t-r/c)}}{r^2} \right]$$

The first term is the radiation field and the second term is the induction field. Hence, for the radiation field only we obtain

$$H_\varphi = \frac{j I_0 e^{j\omega(t-r/c)} \, dl \sin \theta}{2\lambda r}$$

and $$E = \eta H_\varphi = 120\pi H_\varphi$$

where η is the intrinsic impedance of free space.

Appendix C: Loop antenna

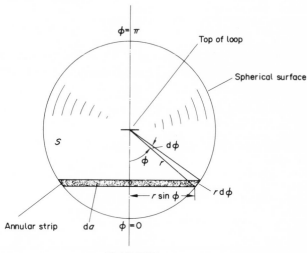

Fig. 55   PLAN VIEW

The power radiated from the loop antenna passes through a spherical shell of radius r. If P_d is the power density at distance r then

$$P_d = E \times H^* \text{ watts/m}^2$$

where $$E = \frac{120\pi^2 I_0 A e^{j\omega(t-r/c)} \sin \varphi}{\lambda^2 r} \quad \text{(Section 2.4)}$$

$$H^* = E^*/120\pi = \frac{\pi I_0 A e^{-j\omega(t-r/c)} \sin \varphi}{\lambda^2 r}$$

or
$$P_d = \frac{120\pi^3 I_0^2 A^2 \sin^2 \varphi}{\lambda^4 r^2}$$

and is independent of the angle θ.

Hence, the *total* average power P radiated is given by

$$P = \int_S \tfrac{1}{2}(E \times H^*)\, da = \int_S \tfrac{1}{2}(P_d \cdot da)$$

where da is an elementary annular area on the shell and integration is over its entire surface S.

For spherical co-ordinates (r, θ, φ) we have from Fig. 55

$$da = 2\pi r \sin \varphi r\, d\varphi = 2\pi r^2 \sin \varphi\, d\varphi$$

Hence
$$P = \int_S \frac{1}{2}\left[\frac{120\pi^3 I_0^2 A^2 \sin^2 \varphi}{\lambda^4 r^2}\right] 2\pi r^2 \sin \varphi\, d\varphi$$

or
$$P = \frac{120\pi^4 I_0^2 A^2}{\lambda^4} \int_0^\pi \sin^3 \varphi\, d\varphi$$

since φ varies from 0 to π over the whole surface S.

Now
$$\int_0^\pi \sin^3 \varphi\, d\varphi = \tfrac{4}{3}$$

Hence
$$P = \frac{120\pi^4 I_0^2 A^2(\tfrac{4}{3})}{\lambda^4}$$

or
$$P = 160\pi^4 I_0^2 (A/\lambda^2)^2$$

Appendix D: Half-wave dipole

Consider two doublets A and B, each of length dz and spaced a distance z apart. The current distribution over the dipole is given by $I_0 \cos(2\pi z/\lambda)$ and radiation from doublet A lags that from doublet B by $z \cos \theta$. Hence, the electric field intensity due to doublet A at a point P distant r is given by

$$dE_\theta = \frac{j60\pi I_0 \cos\left(\dfrac{2\pi z}{\lambda}\right) e^{j\omega[t-(r-z\cos\theta/c)]}}{\lambda r} \sin \theta\, dz$$

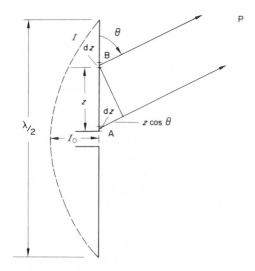

Fig. 56

The total radiation field E_θ at P is given by

$$E_\theta = \int_{-\lambda/4}^{+\lambda/4} j60\pi I_0 \sin\theta e^{j\omega[t-(r-z\cos\theta/c)]} \cos\left(\frac{2\pi z}{\lambda}\right) dz/\lambda \cdot r$$

$$= \frac{j60\pi I_0 \sin\theta e^{j\omega(t-r/c)}}{\lambda \cdot r} \int_{-\lambda/4}^{+\lambda/4} e^{j\omega z\cos\theta/c} \cos\left(\frac{2\pi z}{\lambda}\right) dz$$

Put

$$\beta = \frac{2\pi}{\lambda}$$

$$\alpha = \frac{j\omega z\cos\theta}{c} = j\beta\cos\theta$$

$$E_0 = \frac{j60\pi I_0 \sin\theta}{\lambda r} e^{j\omega(t-r/c)}$$

Hence

$$E_\theta = E_0 \int_{-\lambda/4}^{+\lambda/4} e^{\alpha z} \cos\beta z\, dz$$

Now

$$\int e^{\alpha z} \cos\beta z\, dz = \frac{e^{\alpha z}}{\alpha^2 + \beta^2} [\alpha\cos\beta z + \beta\sin\beta z]$$

Hence $E_\theta = E_0 \left[\dfrac{e^{\alpha z}}{(\alpha^2 + \beta^2)} (\alpha \cos \beta z + \beta \sin \beta z) \right]_{-\lambda/4}^{+\lambda/4}$

$= \dfrac{E_0}{(\alpha^2 + \beta^2)} [e^{\alpha \lambda/4} \{ \alpha \cos (\pi/2) + \beta \sin (\pi/2) \}$

$- e^{-\alpha \lambda/4} \{ \alpha \cos (-\pi/2) + \beta \sin (-\pi/2) \}]$

Now $\alpha \lambda/4 = j\beta \lambda/4 \cos \theta = j\pi/2 \cos \theta$

$\alpha^2 + \beta^2 = -\beta^2 \cos^2 \theta + \beta^2 = \beta^2 \sin^2 \theta$

Hence

$E_\theta = \dfrac{E_0}{\beta^2 \sin^2 \theta} [e^{j\pi/2 \cos \theta} \times \beta + e^{-j\pi/2 \cos \theta} \times \beta]$

or $E_\theta = \dfrac{E_0}{\beta^2 \sin^2 \theta} [2\beta \cos(\pi/2 . \cos \theta)] = \dfrac{\lambda E_0}{\pi \sin^2 \theta} \cos\left(\dfrac{\pi}{2} \cos \theta \right)$

Substituting for E_0 then yields

$$E_\theta = \dfrac{j60I_0 e^{j\omega(t - r/c)}}{r} \dfrac{\cos\left(\dfrac{\pi}{2} \cos \theta \right)}{\sin \theta}$$

Appendix E: Electron collisions and Earth's field

The effect of collisions between the electrons in the ionosphere is to cause further loss of energy between them. The colliding electrons can be regarded as *momentarily sticking* together, giving rise to a frictional force proportional to velocity and to the frequency of collisions v. The expression for the permittivity of the medium becomes

$$\varepsilon = \varepsilon_0 - \dfrac{Ne^2}{m(\omega^2 + v^2)}$$

or $$\varepsilon/\varepsilon_0 = 1 - \dfrac{Ne^2}{m\varepsilon_0(\omega^2 + v^2)}$$

and $$n = \sqrt{1 - \dfrac{Ne^2}{m\varepsilon_0(\omega^2 + v^2)}}$$

Since the negative term is *decreased*, the value of n is increased slightly.

Furthermore, the medium also has a complex conductivity σ given by

$$\sigma = \frac{Ne^2}{m(\omega^2 + v^2)} (v + j\omega)$$

In the presence of the earth's field **B** the ionosphere becomes *aniso-tropic*. A plane wave incident on such a medium is split into two circularly polarised waves rotating in opposite directions. These are called the *ordinary ray* and the *extraordinary ray*. They travel forward with different velocities v_1 and v_2 where

$$v_1 = \frac{c}{\sqrt{1 - \dfrac{\omega_c^2}{(\omega^2 - Be\omega/m)}}}$$

$$v_2 = \frac{c}{\sqrt{1 - \dfrac{\omega_c^2}{(\omega^2 + Be\omega/m)}}}$$

where ω_c is the critical angular frequency obtained in a field-free medium. Hence, the ionosphere can be associated with two different refractive indices. This can cause a rotation of the plane of polarisation and consequent fading of the received signal.

Index